Hydrogen Production, Storage, and Utilization

Hydrogen Production, Storage, and Utilization: Technologies and Applications presents a comprehensive and in-depth exploration of the scientific and engineering principles of hydrogen technology. Written in a technical and scientific manner, using rigorous scientific language and mathematical models to explain principles and applications, the book covers various aspects of hydrogen technology, ranging from fundamental principles of thermodynamics and kinetics to practical applications of hydrogen production, storage, and utilization.

- Includes chapters on the latest advances in hydrogen production, including methods such as steam methane reforming, electrolysis, and biomass gasification.
- Expresses the scientific principles of hydrogen storage, including metal hydrides, carbon-based materials, and liquid carriers.
- Discusses the latest research on fuel cell technologies, including proton-exchange membrane fuel cells, solid oxide fuel cells, and microbial fuel cells.
- Covers hydrogen safety aspects, including risk assessment, safety protocols, and safety standards.
- Explores the challenges and opportunities associated with the deployment of hydrogen technology, including economic viability, environmental impact, and social acceptance.

The book is intended for scientists, engineers, researchers, and graduate students in the fields of chemical engineering, materials science, renewable energy, and sustainability, as well as policymakers and stakeholders interested in the potential of hydrogen as a clean and renewable energy carrier.

Abbas Tcharkhtchi is a distinguished professor at Arts et Métiers ParisTech (ENSAM). He has previously served as the head of the Polymers and Composites Group at ENSAM, the head of the Processing, Mechanic, and Materials Department, and a member of the research council of ENSAM. At present, he is the scientific manager of doctoral training in the PIMM Laboratory at the ENSAM-Campus Paris, responsible for 70 PhD students. He has published over 200 papers and contributed to the publication of several books.

Hamid Reza Vanaei is an associate professor at the Ecole Supérieure d'Ingénieurs Léonard de Vinci (ESILV), Léonard de Vinci University in Paris, France. He obtained his MSc and PhD in Mechanics of Materials from the Arts et Métiers Institute of Technology in Paris and works as an associate researcher at the same institute. With six years of experience in materials science, mechanical engineering, and advanced manufacturing, his primary research focuses on a multidisciplinary approach to optimizing advanced manufacturing techniques.

Albert Lucas is a research engineer in Mechanical and Materials Engineering at ENSAM in Paris. He specializes in rotational molding processes, nondestructive testing (NDT), and the relationships between processes, microstructures, and mechanical properties. He has co-authored several scientific papers and contributed to patents related to polymer hydrogen tank development. He is currently in the final year of his PhD preparation.

Sedigheh Farzaneh is the Director of P4Tech, an engineering company, in France. She has been a lecturer and researcher at ENSAM, Léonard de Vinci University, and ISIPCA in France for over 30 years. She holds a PhD in Life Sciences from Pierre and Marie Curie University (Paris VI), specializing in areas such as polymers, biopolymers, biomass, and hydrogen technologies. Dr. Farzaneh has led research projects, supervised numerous PhD and Master students, and published over 30 scientific articles.

Emerging Materials and Technologies

Series Editor: Boris I. Kharissov

The *Emerging Materials and Technologies* series is devoted to highlighting publications centered on emerging advanced materials and novel technologies. Attention is paid to those newly discovered or applied materials with potential to solve pressing societal problems and improve quality of life, corresponding to environmental protection, medicine, communications, energy, transportation, advanced manufacturing, and related areas.

The series takes into account that, under present strong demands for energy, material, and cost savings, as well as heavy contamination problems and worldwide pandemic conditions, the area of emerging materials and related scalable technologies is a highly interdisciplinary field, with the need for researchers, professionals, and academics across the spectrum of engineering and technological disciplines. The main objective of this book series is to attract more attention to these materials and technologies and invite conversation among the international R&D community.

2D Semiconductors for Environmental Remediation
Edited by Honey John, Nisha T Padmanabhan, Sona Stanly and Jith C Janardhanan

Materials from Natural Sources: Structure, Properties, and Applications
Edited by Ramesh Gardas, Neha Patni, and Amita Chaudhary

Dielectric Materials for Capacitive Energy Storage
Edited By Haibo Zhang and Hua Tan

Multifunctional Coordination Materials for Green Energy Technologies
Edited by Ghulam Yasin, Anuj Kumar, Sajjad Ali, Tuan Anh Nguyen,
and Saira Ajmal

Advancements in Nanomaterials for Energy Conversion and Storage
Edited by Piyush Kumar Sonkar and Vellaichamy Ganesan

2D Materials-Based Sensors: Technology and Applications
Vinod Kumar Khanna

Metal Organic Framework Derived Materials: Design Strategies and Applications
Gomathi Nageswaran, Varsha M V, Arun Kumar Rajasekaran and M Shashank Rao

Hydrogen Production, Storage, and Utilization: Technologies and Applications
Abbas Tcharkhtchi, Hamid Reza Vanaei, Albert Lucas, and Sedigheh Farzaneh

For more information about this series, please visit: www.routledge.com/Emerging-Materials-and-Technologies/book-series/CRCEMT

Hydrogen Production, Storage, and Utilization
Technologies and Applications

Abbas Tcharkhtchi

Hamid Reza Vanaei

Albert Lucas

Sedigheh Farzaneh

CRC Press
Taylor & Francis Group
Boca Raton London New York

CRC Press is an imprint of the
Taylor & Francis Group, an **informa** business

Designed cover image: © Shutterstock, Retouch man

First edition published 2025
by CRC Press
2385 NW Executive Center Drive, Suite 320, Boca Raton FL 33431

and by CRC Press
4 Park Square, Milton Park, Abingdon, Oxon, OX14 4RN

CRC Press is an imprint of Taylor & Francis Group, LLC

Library of Congress Cataloging-in-Publication Data
Names: Tcharkhtchi, Abbas, author. | Vanaei, Hamid Reza, author. | Lucas, A. (Albert), author. | Farzaneh, Sedigheh, author.
Title: Hydrogen production, storage, and utilization : technologies and applications / by Abbas Tcharkhtchi, Hamidreza Vanaei, Albert Lucas, and Sedigheh Farzaneh.
Description: First edition. | Boca Raton : CRC Press, 2025. | Series: Emerging materials and technologies | Includes bibliographical references and index. | Identifiers: LCCN 2024034541 (print) | LCCN 2024034542 (ebook) | ISBN 9781032713038 (hbk) | ISBN 9781032718439 (pbk) | ISBN 9781032718453 (ebk)
Subjects: LCSH: Hydrogen as fuel. | Hydrogen--Storage. | Renewable energy sources.
Classification: LCC TP359.H8 T395 2025 (print) | LCC TP359.H8 (ebook) | DDC 665.8/1--dc23/eng/20241009
LC record available at https://lccn.loc.gov/2024034541
LC ebook record available at https://lccn.loc.gov/2024034542

ISBN: 978-1-032-71303-8 (hbk)
ISBN: 978-1-032-71843-9 (pbk)
ISBN: 978-1-032-71845-3 (ebk)

DOI: 10.1201/9781032718453

Typeset in Times
by SPi Technologies India Pvt Ltd (Straive)

Contents

Preface

In the annals of scientific and technological evolution, few elements have captured the imagination and promise of future sustainability as profoundly as hydrogen. This book, "Hydrogen Production, Storage, and Utilization: Technologies and Applications," emerges from a confluence of rigorous research, innovative technological strides, and a pressing global need for sustainable energy solutions. Our endeavor is to present a comprehensive and nuanced exploration of hydrogen, encapsulating its potential to revolutionize the energy landscape.

Hydrogen, the universe's most abundant element, holds unparalleled promise in addressing the dual challenges of energy security and environmental sustainability. Its unique properties as a clean energy carrier, capable of being produced from various sources and yielding only water as a byproduct, position it as a cornerstone for future energy systems. This potential, however, is met with significant challenges in production, storage, and utilization, necessitating a detailed and scholarly examination.

The inception of this book is rooted in the urgent call for sustainable energy solutions that can mitigate the detrimental impacts of fossil fuels. The global community's commitment to reducing greenhouse gas emissions and transitioning to low-carbon economies underscores the critical role hydrogen is poised to play. This book aims to serve as both a foundational text and a forward-looking treatise, providing readers with a robust understanding of hydrogen technologies while also charting the path toward future innovations.

The structure of this book is meticulously crafted to guide the reader through the multifaceted dimensions of hydrogen technology. Beginning with an introductory chapter that sets the stage, the book presents the elemental and molecular characteristics of hydrogen, elucidating why it is considered the fuel of the future.

Subsequent chapters are dedicated to the diverse methodologies of hydrogen production. These include traditional processes such as steam methane reforming (SMR) coupled with carbon capture and storage (CCS), and cutting-edge techniques like water electrolysis powered by renewable energy sources. Each method is scrutinized for its efficiency, economic viability, and environmental impact, providing a holistic view of current and emerging production technologies.

The intricacies of hydrogen storage are explored in depth, acknowledging the critical role storage plays in the practical application of hydrogen as an energy carrier. The book examines high-pressure gas storage, cryogenic liquid storage, and advanced materials-based storage solutions such as metal hydrides and carbon nanotubes. These discussions are framed within the context of technological advancements aimed at overcoming the inherent challenges of hydrogen's low volumetric energy density and the need for safe, efficient storage systems.

The utilization of hydrogen spans a wide array of applications, each of which is meticulously covered in this volume. From its use in fuel cells for transportation and stationary power generation to its role in industrial processes and as a means of storing and balancing renewable energy, the book provides a comprehensive overview. Particular attention is given to the environmental benefits of hydrogen applications, including significant reductions in greenhouse gas emissions and improvements in air quality.

The book also addresses the broader societal and policy dimensions of hydrogen technology. This includes discussions on safety standards, regulatory frameworks, and the critical importance of public awareness and education in fostering a hydrogen economy. The role of government policies and international collaborations is highlighted as pivotal in accelerating the adoption and integration of hydrogen technologies into mainstream energy systems.

As we embark on this exploration of hydrogen, it is imperative to recognize the collective effort and interdisciplinary collaboration that underpin the advancements in this field. The insights and knowledge presented in this book are the result of contributions from experts across various domains, unified by a shared vision of a sustainable and resilient energy future.

We invite you, the reader, to engage with the material with an open mind and a forward-looking perspective. The journey toward a hydrogen-powered world is laden with challenges but equally rich in opportunities for innovation and growth. By embracing hydrogen's potential, we can pave the way for a cleaner, greener future, ensuring energy security and environmental stewardship for generations to come.

In closing, we extend our gratitude to the scientific community, industry stakeholders, policymakers, and educators whose relentless pursuit of excellence and dedication to sustainability have made this book possible. Together, let us embark on the path to a hydrogen economy, forging a legacy of innovation and environmental harmony.

Authors

1 Introduction

1.1 HYDROGEN: ELEMENT OF THE FUTURE IN SUSTAINABLE ENERGY SOLUTIONS

Hydrogen is the most abundant in the universe, comprising about 92% of its total composition. On Earth, hydrogen primarily exists in water and hydrocarbon molecules. Dihydrogen, a colorless, odorless, and nontoxic gas, has a density of 0.08 g/L at room temperature and atmospheric pressure, with melting and boiling points of −259.3°C and −252.9°C, respectively. Despite its abundance in the universe, hydrogen is rare in Earth's atmosphere, constituting only about 1 ppm by volume. Discovered by Henry Cavendish in 1776 and named by Antoine Lavoisier, hydrogen plays a crucial role in various industries, including ammonia (NH_3) and methanol (CH_3OH) production, petroleum refining, food processing, and aerospace. Hydrogen's unique properties make it an optimal energy choice, capable of being produced and converted into electricity with high efficiency and minimal emissions when used in fuel cells. As a carbon-free fuel, it produces only water during combustion and very low emissions of nitrogen oxides. Hydrogen's high calorific value, although presenting challenges in volumetric density, can be addressed through compression, making it economically competitive for mobile and transportation applications (Figure 1.1). The main challenge is reducing the size of hydrogen tanks for vehicles to store significant quantities in small volumes. Hydrogen's potential as a sustainable energy solution is vast, with its role in reducing carbon emissions and providing a renewable energy source, underscoring its importance in the future of energy.

Figure 1.2 outlines the evolutionary path of hydrogen utilization across a diverse range of applications. From fuel cells powering vehicles and stationary applications to stoves, turbines, internal combustion engines (ICEs), and industrial boilers, hydrogen demonstrates its versatility across different sectors. The true potential of hydrogen electrolysis and fuel cell technologies becomes apparent when they are extensively deployed and integrated into numerous applications. Scaling up these technologies enhances their economic viability and efficiency and opens up new avenues for advancing technology. This scalability drives down costs and promotes innovation and broader adoption. In broader contexts, hydrogen assumes critical roles, such as balancing the electric grid by storing excess renewable energy and facilitating its use during peak demand periods. Moreover, hydrogen is pivotal in decarbonizing the transportation sector, offering zero-emission solutions for cars, trucks, buses, and maritime vessels. Its importance extends further into essential industrial processes, including chemical manufacturing and refining, where hydrogen serves as a clean and efficient feedstock or energy carrier [1, 2]. Hence, hydrogen's multifaceted applications underscore its potential as a cornerstone of future energy systems. Its integration across sectors promises to mitigate carbon emissions, drive sustainable economic growth, and foster technological innovation.

DOI: 10.1201/9781032718453-1

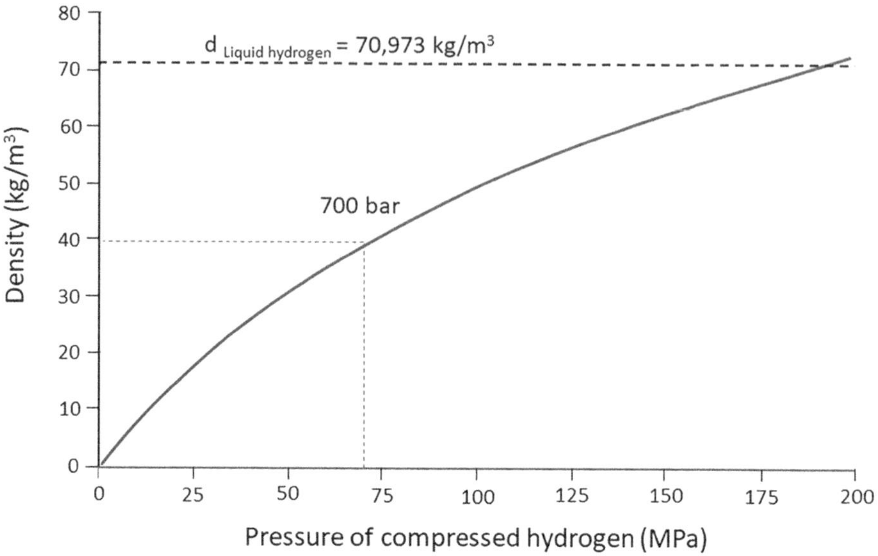

FIGURE 1.1   Density of compressed hydrogen versus pressure.

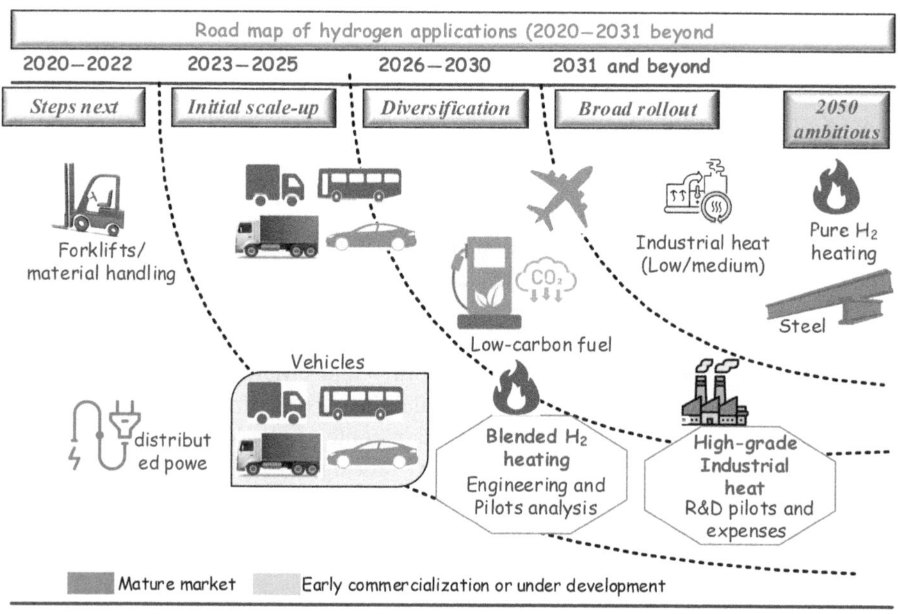

FIGURE 1.2   Hydrogen applications roadmap (2020–2031 beyond).

(Adapted from ref. [3]).

1.1.1 HYDROGEN ATOMS

The first element of the periodic table of elements, hydrogen is the lightest element in the universe, with an atom weighing approximately 1.0078 grams per mole (Figure 1.3). It is also the primary component of the universe, representing about 92% of its composition. On Earth, it is mainly associated with water molecules and hydrocarbons, forming bonds exclusively with other atoms. The gas derived from dihydrogen molecules is characterized by its lack of color, odor, and toxicity, displaying a density of 0.08 g/L at room temperature and atmospheric pressure. Its melting and boiling points are −259.3°C and −252.9°C, respectively. The dihydrogen molecule exists naturally, primarily in oceanic trenches, but so far, no technology has allowed its exploitation. It adopts two distinct isomeric forms: 75% ortho-hydrogen, characterized by parallel spins of the protons, and 25% para-hydrogen, where the spins are opposite.

It is necessary to specify that hydrogen, which is the first element of Mendeleev's table, as said before, is known for its abundance in many materials. This property can be advantageously used to store hydrogen for powering generators such as fuel cells or for making electrodes for electrochemical batteries. Henry Cavendish discovered hydrogen as a distinct entity in 1776, but it was Antoine Lavoisier who gave it its name. Although hydrogen is the most abundant element in the universe, constituting 75% of its mass and 90% of its number of atoms, it is extremely rare in the Earth's atmosphere, at only about 1 ppm by volume.

The main methods of hydrogen production include steam reforming of light hydrocarbons, partial oxidation (PO) of various hydrocarbons (light, heavy, and asphalts), as well as coal and biomass. Currently, electrolysis of water accounts for only a tiny fraction of hydrogen production, mainly due to its high cost attributable to significant energy consumption. Hydrogen plays a pivotal industrial role, with an estimated global annual demand of approximately 110 million tons (Mt). It is widely used in various industrial processes such as the production of NH_3 and CH_3OH. Notably, the synthesis of NH_3 alone accounts for a consumption of 20 Mt per year. In the petroleum industry, hydrogen is essential for hydrodesulfurization, hydrogenation of unsaturated hydrocarbons, and hydrocracking of distillates. Additionally, it is employed in the food industry for the hydrogenation of oils and other fatty substances. Liquid hydrogen is also utilized in the aerospace field for spacecraft propulsion.

Hydrogen possesses unique properties that make it an optimal energy choice [4]. These characteristics include: (a) its ability to be produced and converted into electricity with high efficiencies; (b) its status as a renewable or recyclable fuel within the hydrogen cycle; (c) its potential to be stored in gaseous form (suitable for large-scale storage systems) or liquid form (for aerial and space applications); and (d) its environmental friendliness, as its use in a fuel cell generates fewer pollutants and greenhouse gases (GHGs) than other fuels [5].

Current hydrocarbon fuels emit approximately the same amount of carbon dioxide (CO_2) per unit of power. Even hydrocarbons with a high percentage of hydrogen show less than a 20% difference in CO_2 emission rates between C_2H_2 and C_2H_4. In contrast, hydrogen stands out as a carbon-free fuel. Unlike hydrocarbons, the

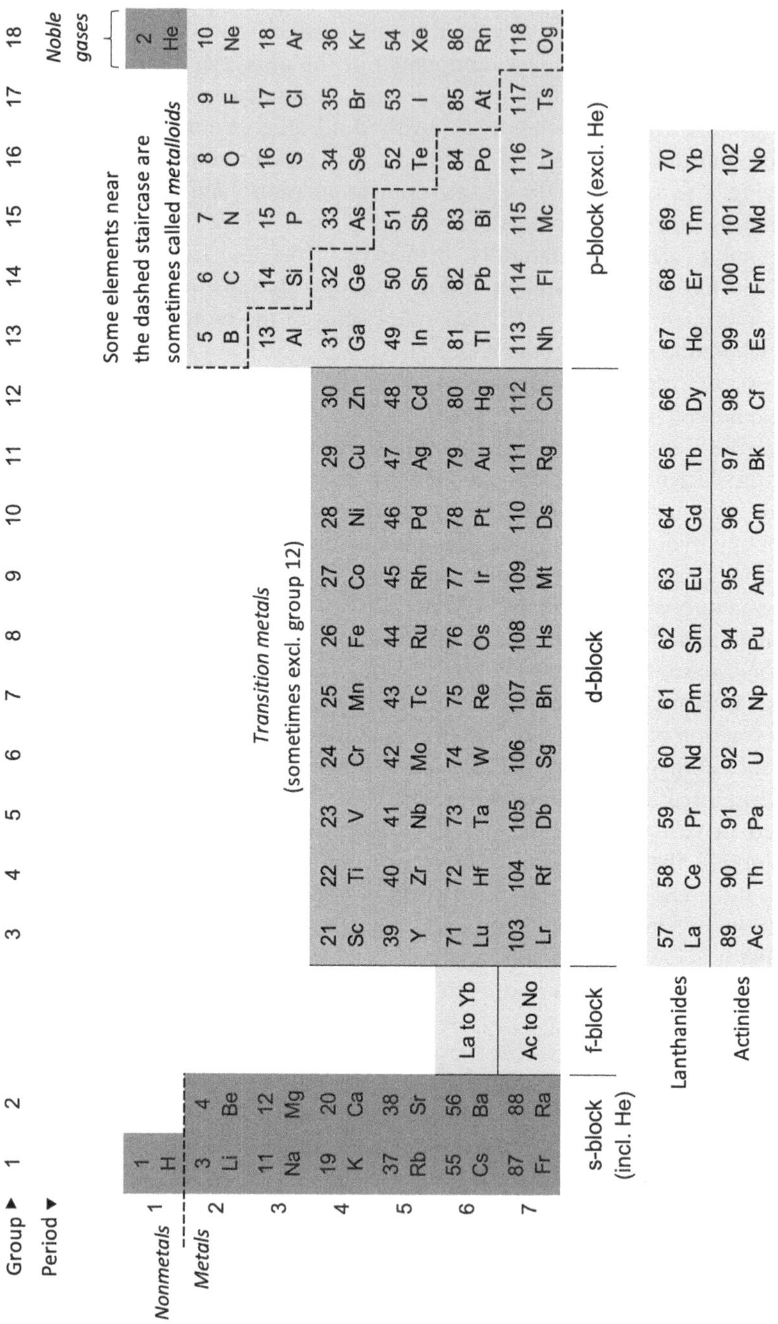
Group ▸ 1 2 3 4 5 6 7 8 9 10 11 12 13 14 15 16 17 18
Period ▾
Nonmetals
Metals
Noble gases
Some elements near the dashed staircase are sometimes called metalloids
1 1 H 2 He
2 3 Li 4 Be 5 B 6 C 7 N 8 O 9 F 10 Ne
3 11 Na 12 Mg 13 Al 14 Si 15 P 16 S 17 Cl 18 Ar
4 19 K 20 Ca 21 Sc 22 Ti 23 V 24 Cr 25 Mn 26 Fe 27 Co 28 Ni 29 Cu 30 Zn 31 Ga 32 Ge 33 As 34 Se 35 Br 36 Kr
5 37 Rb 38 Sr 39 Y 40 Zr 41 Nb 42 Mo 43 Tc 44 Ru 45 Rh 46 Pd 47 Ag 48 Cd 49 In 50 Sn 51 Sb 52 Te 53 I 54 Xe
6 55 Cs 56 Ba La to Yb 71 Lu 72 Hf 73 Ta 74 W 75 Re 76 Os 77 Ir 78 Pt 79 Au 80 Hg 81 Tl 82 Pb 83 Bi 84 Po 85 At 86 Rn
7 87 Fr 88 Ra Ac to No 103 Lr 104 Rf 105 Db 106 Sg 107 Bh 108 Hs 109 Mt 110 Ds 111 Rg 112 Cn 113 Nh 114 Fl 115 Mc 116 Lv 117 Ts 118 Og
Transition metals
(sometimes excl. group 12)
s-block (incl. He)
f-block
d-block
p-block (excl. He)
Lanthanides 57 La 58 Ce 59 Pr 60 Nd 61 Pm 62 Sm 63 Eu 64 Gd 65 Tb 66 Dy 67 Ho 68 Er 69 Tm 70 Yb
Actinides 89 Ac 90 Th 91 Pa 92 U 93 Np 94 Pu 95 Am 96 Cm 97 Bk 98 Cf 99 Es 100 Fm 101 Md 102 No

FIGURE 1.3 Periodic table of chemical elements.

(Adapted from Wikipedia.)

combustion of hydrogen does not generate any toxic molecules such as carbon monoxide (CO), sulfur oxides, organic acids, and CO_2. Hydrogen combustion mainly produces water, with very low emissions of nitrogen oxides (NO_x) if the hydrogen used contains impurities.

Hydrogen has the highest calorific value. On an equivalent mass basis, hydrogen combustion provides 2.8 times more energy than gasoline. However, its low volumetric density poses a challenge. By compressing hydrogen to increase the mass stored per unit volume, it can achieve performance comparable to gasoline in terms of volumetric energy density. Storage is crucial to make hydrogen economically competitive in mobile and transportation applications. It must meet numerous requirements of automotive manufacturers and users.

A vehicle running on hydrogen can utilize it in two ways: either by burning it in an ICE with air oxygen or by electrochemically consuming it with air oxygen in a fuel cell. In practice, a modern vehicle available on the market, offering a range of 400 kilometers, requires approximately 24 kilograms of petroleum in an ICE. To achieve similar performance, it would require 8 kilograms of hydrogen for the ICE version or 4 kilograms of hydrogen in the case of a fuel cell-powered electric vehicle. In the latter scenario, hydrogen would occupy a volume of 45 cubic meters at atmospheric pressure and room temperature. Therefore, the main challenge is to reduce the size of the hydrogen tank. If the pressure increases to 700 bars, it is possible to store 4 kilograms of hydrogen gas in a tank with a volume of 65 liters.

Unlike naturally occurring hydrocarbons, hydrogen is artificially produced by humans, which incurs additional costs. Currently, this results in a final price approximately three times higher than petroleum products. Additionally, the method of hydrogen storage as fuel must not significantly add to the costs. The need to capture and release hydrogen reversibly excludes the use of covalent hydrocarbon compounds as storage materials, as hydrogen is released at temperatures exceeding 800°C. Interesting storage methods include compression, liquefaction, metal hydrides, and physisorption.

The choice of hydrogen as an energy source can be justified by various criteria and parameters, each corresponding to specific advantages of hydrogen. First, the energy density of hydrogen is a major asset. On a gravimetric basis, hydrogen has a very high energy density per unit of mass, approximately 120 MJ/kg, making it a light and efficient fuel, particularly advantageous for applications requiring lightweight solutions, such as aviation and heavy vehicles. Although gaseous hydrogen has a low volumetric density, technologies such as high-pressure compression or liquefaction allow for the storage of large amounts of energy in reasonable volumes.

In terms of emissions and the environment, hydrogen offers significant advantages. When used in fuel cells, it produces only water as a by-product, making it a clean option in terms of local emissions. Furthermore, when produced from renewable sources, such as electrolysis using green electricity, hydrogen can significantly reduce GHG emissions over its entire lifecycle, contributing to the fight against climate change.

The flexibility and storage of energy are also critical criteria in favor of hydrogen. It can be stored for long periods, unlike electricity in batteries, which can discharge over time. Moreover, hydrogen can be used in various sectors, including mobility (cars, trucks, trains, and planes), industry (steel and chemicals), and domestic heating, making it a versatile energy source.

In terms of energy independence and security, hydrogen allows for the diversification of energy sources. It can be produced from various sources, including renewables, natural gas, and even waste, thus contributing to energy security by diversifying supply sources. Local production of hydrogen can also reduce dependence on fossil fuel imports, enhancing the energy independence of countries.

Technological development plays a crucial role in the choice of hydrogen. Technologies for the production, storage, and utilization of hydrogen are rapidly developing, with costs decreasing and efficiencies increasing. Additionally, the transition to hydrogen can utilize much of the existing infrastructure for natural gas, with adaptations, thus facilitating its deployment.

Finally, the economic aspect and market for hydrogen are factors to consider. The development of the hydrogen sector can create jobs in production, distribution, and new technologies. As technologies develop and large-scale production intensifies, the costs of hydrogen production should decrease, making this energy source more competitive.

1.1.2 DIFFERENT TYPES OF HYDROGEN

Green hydrogen and other types of hydrogen are terms used to describe different methods of hydrogen production and their associated energy sources (Figure 1.4). These different forms of hydrogen have different implications in terms of carbon footprint, production costs, and sustainability.

Gray hydrogen is a term that specifically refers to hydrogen produced by the process of Steam Methane Reforming (SMR). This method involves reacting natural gas, which is primarily methane (CH_4), with steam (H_2O) under high-pressure and temperature. The chemical reaction produces hydrogen (H_2) and CO_2 as by-products. Despite being the most prevalent form of hydrogen production worldwide due to its cost-effectiveness and established technology, the production of gray hydrogen poses significant environmental concerns. The process of generating gray hydrogen is energy intensive and results in the release of substantial amounts of CO_2 into the atmosphere, contributing to GHG emissions that drive climate change. For every ton of hydrogen produced, approximately 9 to 12 tons of CO_2 are emitted, depending on the efficiency of the technology used and the composition of the natural gas. This carbon footprint is a major drawback, especially when compared to "green" hydrogen, which is produced using renewable energy sources and water electrolysis, resulting in minimal to no emissions of GHGs.

The reliance on natural gas as a feedstock also ties gray hydrogen production to the availability and price volatility of fossil fuel resources, adding an additional layer of complexity to its sustainability profile. Furthermore, as global efforts to combat climate change intensify, the environmental impact of gray hydrogen production has come under increasing scrutiny, leading to a push for alternative, cleaner methods of hydrogen production that can provide the benefits of hydrogen as a clean energy carrier without the associated GHG emissions.

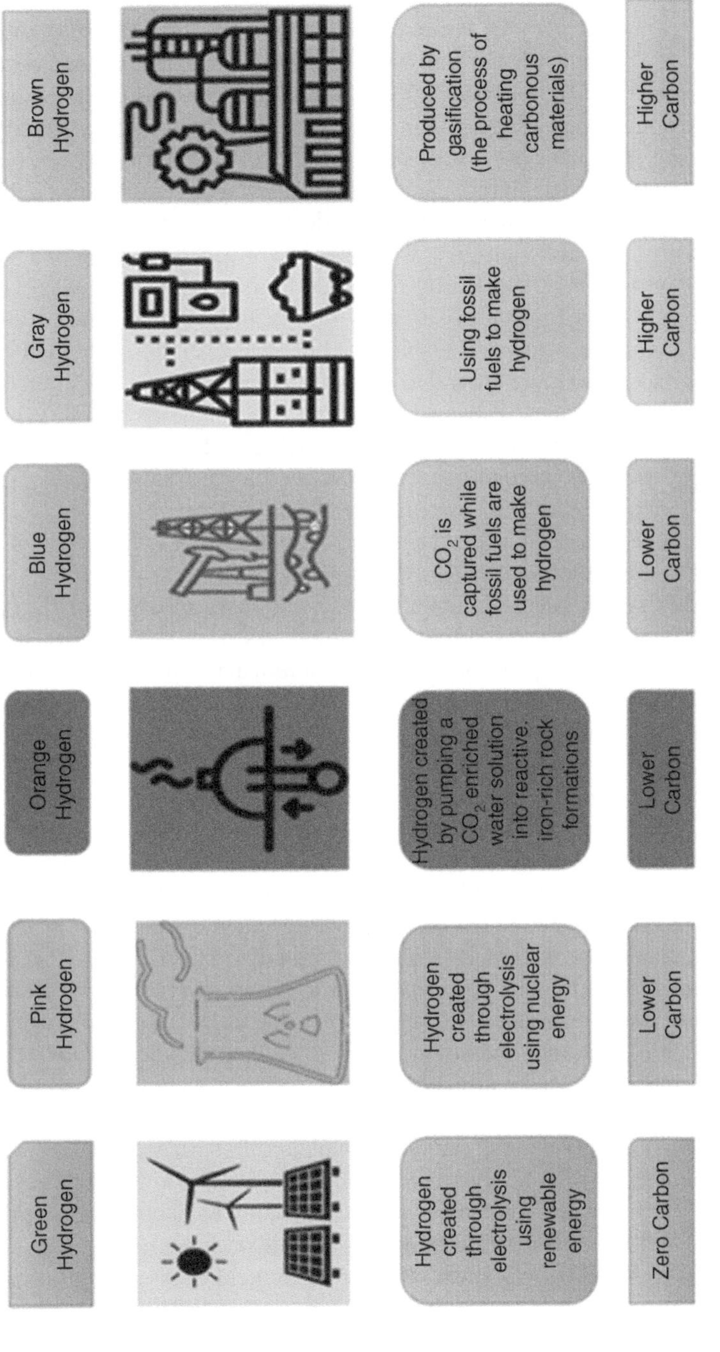

FIGURE 1.4 The color of hydrogen. (Adapted from ref. [6].)

Blue Hydrogen represents a more environmentally conscious approach to hydrogen production compared to its counterpart, gray hydrogen. It is produced through a similar initial process, typically using SMR, where natural gas is reacted with steam under high temperatures and pressure to produce hydrogen and CO_2 as by-products. However, the key differentiator for blue hydrogen lies in the subsequent steps taken to mitigate the environmental impact associated with CO_2 emissions. In the production of blue hydrogen, the CO_2 generated during the SMR process is captured using various technologies such as pre-combustion capture, oxy-fuel combustion, or post-combustion capture. Once captured, the CO_2 is then transported, often through pipelines, and securely stored in underground geological formations, such as depleted oil and gas fields or deep saline aquifers. This process of carbon capture and storage (CCS) effectively prevents the majority of the CO_2 produced during hydrogen production from reaching the atmosphere, thereby significantly reducing the GHG emissions associated with the process. The ability of blue hydrogen to substantially lower CO_2 emissions makes it a more attractive option for supporting the transition toward a low-carbon energy system. However, the feasibility and sustainability of blue hydrogen production depend on several factors, including the efficiency of the capture technology, the integrity and capacity of the geological storage sites, and the overall energy and economic costs associated with the CCS process. It's important to note that while blue hydrogen offers a cleaner alternative to gray hydrogen, it is not completely emissions-free. Some CO_2 emissions may still occur during the production process due to incomplete capture efficiency, and there are also emissions associated with extracting and transporting the natural gas used as feedstock. Despite these challenges, blue hydrogen is considered a significant step forward in the effort to decarbonize the energy sector, providing a pathway to reduce emissions from hydrogen production until green hydrogen, produced entirely from renewable energy and water, becomes more widely viable and cost-effective.

Pink Hydrogen: Pink hydrogen is a promising form of clean energy that combines the benefits of hydrogen fuel with the low-carbon footprint of nuclear power. Unlike traditional hydrogen production methods, which often rely on fossil fuels, pink hydrogen is generated through electrolysis using electricity from nuclear reactors. This process splits water molecules into hydrogen and oxygen without emitting GHGs, making it an environmentally friendly alternative. The use of nuclear energy ensures a stable and continuous power supply for hydrogen production, addressing the intermittency issues associated with renewable sources like wind and solar power. The development of pink hydrogen is seen as a crucial step toward achieving global decarbonization goals. It can be utilized in various sectors, including transportation, industry, and energy storage, providing a versatile and sustainable solution to reduce carbon emissions. Countries with established nuclear infrastructure can leverage this technology to enhance their energy security and transition to a low-carbon economy. Additionally, pink hydrogen can play a significant role in balancing energy grids by storing excess electricity generated during low-demand periods and releasing it when needed. As research and development (R&D) in this field progress, pink hydrogen has the potential to become a key component of the global clean energy landscape.

Orange Hydrogen: Orange hydrogen refers to hydrogen produced through the process of biomass gasification, a method that involves converting organic materials such as agricultural waste, wood chips, or other biomass into hydrogen gas. This process is achieved by exposing the biomass to high temperatures in the presence of a controlled amount of oxygen or steam, resulting in the production of syngas, which primarily consists of hydrogen, CO, and CO_2. The hydrogen is then separated and purified for various uses. Orange hydrogen stands out as a renewable and sustainable energy source, as it utilizes biomass, which is abundantly available and can be replenished through natural processes, making it an attractive option for reducing dependence on fossil fuels. One of the significant advantages of orange hydrogen is its potential for carbon neutrality or even carbon negativity. When biomass is grown, it absorbs CO_2 from the atmosphere, which is then released during the gasification process. However, if the CO_2 produced is captured and stored through CCS technologies, the overall process can result in a net reduction of atmospheric CO_2. This makes orange hydrogen a compelling option for mitigating climate change while providing a reliable source of clean energy. Moreover, utilizing agricultural and forestry waste for hydrogen production can help manage waste and reduce environmental pollution. The versatility of orange hydrogen allows it to be integrated into various sectors, including transportation, industry, and power generation. In the transportation sector, it can be used as a fuel for hydrogen fuel cell vehicles, providing a clean alternative to conventional gasoline and diesel engines. In industrial applications, orange hydrogen can replace fossil fuels in processes that require high temperatures, such as steel production and chemical manufacturing. Additionally, it can be used for power generation in fuel cells, contributing to a diversified and resilient energy system. As technology advances and the infrastructure for biomass gasification and hydrogen utilization expands, orange hydrogen has the potential to play a crucial role in the transition to a sustainable and low-carbon energy future.

Green Hydrogen: Green hydrogen represents the pinnacle of clean energy within the context of hydrogen production technologies due to its virtually zero emissions profile throughout the production process. It is produced through a method known as electrolysis, a process in which electrical current is used to split water (H_2O) into its constituent elements—hydrogen (H_2) and oxygen (O_2). The defining characteristic of green hydrogen is the source of the electricity used in this process: it must be derived entirely from renewable energy sources, such as solar, wind, or hydroelectric power. Electrolysis involves passing an electrical current through water, which has been slightly acidified or alkalized to improve conductivity, using an electrolyzer. This system consists of two electrodes (an anode and a cathode) submerged in the water, separated by an electrolyte. When the electrical current flows through the water, hydrogen gas bubbles up at the cathode, and oxygen gas at the anode, both of which are then collected. The environmental impact of green hydrogen production is minimal. Since the electricity used comes from renewable sources, the entire process can be carbon-neutral, assuming the lifecycle emissions associated with the production and maintenance of the renewable energy infrastructure are also taken into account. This makes green hydrogen a highly sought-after solution in efforts to decarbonize sectors that are challenging to electrify directly, such as heavy industry (e.g., steel and chemical production) and heavy transport (e.g., shipping and aviation). However, the scalability of green hydrogen production faces several challenges. The cost of electrolyzers and the current price of electricity from renewable sources can make green hydrogen more expensive than hydrogen produced from fossil fuels (gray or blue hydrogen). Additionally, the availability of renewable energy sources can be intermittent, which might affect the consistent production of green hydrogen unless paired with energy storage solutions or a more flexible grid. Despite these challenges, advancements in electrolyzer technology, decreasing costs of renewable energy, and growing policy support are expected to enhance the viability and cost-competitiveness of green hydrogen in the near future. As such, green hydrogen is considered a crucial component of the global strategy to achieve a net-zero carbon economy, offering a sustainable and clean solution for energy storage, transportation, and industrial processes.

Brown Hydrogen: Brown hydrogen is hydrogen produced through the process of coal gasification, a method that involves reacting coal with oxygen and steam at high temperatures to produce a mixture of gases known as syngas, which primarily contains hydrogen, CO, and CO_2. This process is well-established and has been used for decades to produce industrial hydrogen. However, it is highly carbon-intensive because it relies on coal, one of the most polluting fossil fuels. As a result, brown hydrogen production releases significant amounts of CO_2 and other GHGs into the atmosphere, contributing to global warming and climate change. Despite its environmental drawbacks, brown hydrogen remains economically attractive in regions with abundant coal reserves and limited access to cleaner energy sources. It provides a way to utilize domestic coal resources and can serve as a transitional technology while cleaner hydrogen production methods, such as green or blue hydrogen, are developed and scaled up. In some cases, brown hydrogen facilities can be retrofitted with CCS technologies to reduce their carbon footprint. However, the feasibility and economic viability of CCS on a large scale remain uncertain, and the overall sustainability of brown hydrogen is still a matter of debate. The future of brown hydrogen is closely tied to advancements in carbon capture technologies and the broader energy transition toward decarbonization. While it can currently provide a bridge solution in the shift away from fossil fuels, its long-term role is likely to diminish as renewable energy sources and low-carbon hydrogen production methods become more cost-effective and widespread. Policymakers and industry leaders are increasingly focusing on reducing emissions and investing in cleaner alternatives to meet climate goals. Consequently, the emphasis on brown hydrogen is expected to decline, giving way to more sustainable forms of hydrogen that align better with global efforts to combat climate change and promote environmental sustainability.

The World Bank Group is implementing a range of initiatives in Brazil, Panama, Costa Rica, Colombia, and Chile aimed at establishing green hydrogen as a viable fuel and promoting its use for energy storage. Similarly, countries such as Australia, India, China, Japan, Bangladesh, and Germany are planning their energy transitions with a strong emphasis on green hydrogen. Figure 1.5 illustrates the global potential for green hydrogen production by region. Sub-Saharan Africa leads with the highest potential at 2715 EJ, followed by the Middle East and North Africa at 2023 EJ. North America (1314 EJ), Oceania (1272 EJ), South America (1114 EJ), the rest of Asia (684 EJ), Northeast Asia (212 EJ), and Southeast Asia (64 EJ) also show significant potential.

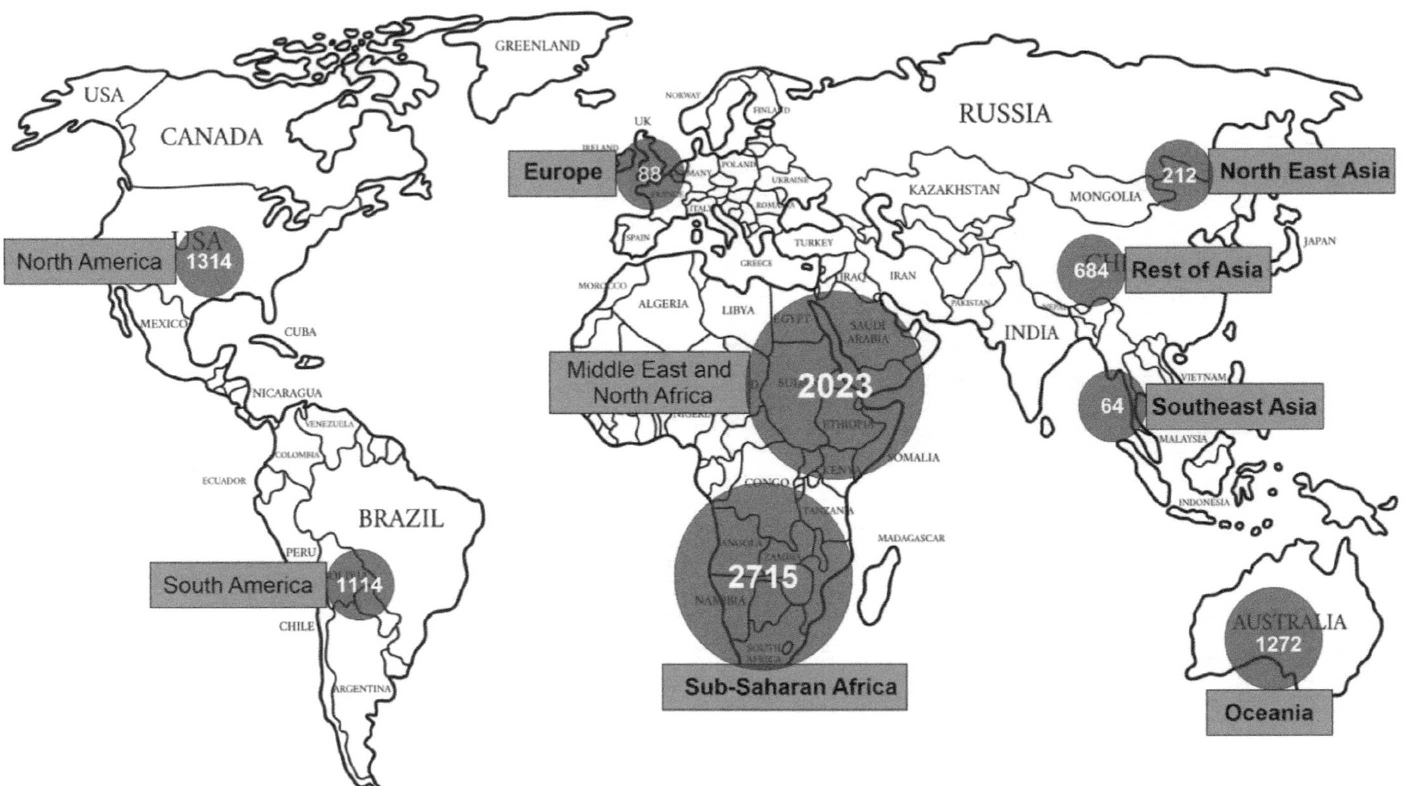

FIGURE 1.5 Global green hydrogen production (in EJ) potential by region.

(Adapted from ref. [7].)

1.1.3 Clean and Efficient Hydrogen Production

Hydrogen production methods encompass a spectrum of scientific principles and technologies aimed at generating hydrogen gas with varying degrees of efficiency and environmental impact. Water electrolysis involves the splitting of water molecules (H_2O) into hydrogen (H_2) and oxygen (O_2) using an electric current. This process occurs at electrodes submerged in an electrolyte solution. The cathode attracts positively charged hydrogen ions (protons), causing them to gain electrons and form hydrogen gas. Simultaneously, the anode attracts negatively charged hydroxide ions, which lose electrons to form oxygen gas. The efficiency of water electrolysis depends on factors such as the type of electrolyte used, the design of the electrolysis cell, and the source of electricity. Utilizing renewable energy sources like solar or wind power enhances the environmental sustainability of this method.

Methane reforming, specifically SMR, involves the reaction of CH_4 with steam (H_2O) at high temperatures (>700°C) and pressures, typically using a catalyst such as nickel. This process produces hydrogen gas and CO. The resulting mixture, known as synthesis gas or syngas, can undergo further processing to increase the hydrogen yield while reducing the CO content. Although SMR is a well-established industrial process, it generates CO_2 emissions due to the conversion of CH_4, a potent GHG. Implementing CCS technologies can mitigate these emissions, making SMR more environmentally sustainable.

Biomass conversion methods involve the transformation of organic materials, such as agricultural residues, forestry waste, or dedicated energy crops, into hydrogen gas. Processes like gasification and fermentation are commonly utilized. Gasification converts biomass into a mixture of hydrogen, CO, CO_2, and trace gases by subjecting it to high temperatures (>700°C) in a controlled environment with limited oxygen. Fermentation employs microorganisms to break down organic substrates into hydrogen, typically in anaerobic conditions. Biomass-based hydrogen production offers advantages in terms of utilizing renewable feedstocks and reducing reliance on fossil fuels. Biological electrolysis, also known as microbial electrolysis or microbial electrolysis cell (MEC), leverages microbial activity to produce hydrogen from organic substrates. In MECs, microorganisms oxidize organic compounds, releasing electrons that are captured at the anode. Meanwhile, protons migrate through a proton exchange membrane (PEM) to the cathode, where they combine with electrons to form hydrogen gas. This process operates at ambient temperatures and pressures, offering potential energy savings compared to traditional electrolysis. While still under development, biological electrolysis holds promise as a sustainable and renewable method for hydrogen production. These methods represent ongoing areas of R&D aimed at advancing the production of hydrogen as a clean and sustainable energy carrier. Each method presents distinct challenges and opportunities for improving efficiency, reducing costs, and minimizing environmental impacts.

1.1.4 Reducing Hydrogen Production Costs

Reducing the production costs of green hydrogen, derived from renewable energy sources, is pivotal for enhancing the competitiveness and widespread adoption of this environmentally sustainable energy carrier. Various strategies can be employed to achieve this objective, leveraging advancements in scientific understanding and technological innovation. Continuous advancements in water electrolysis technologies play a critical role in cost reduction. Improving the efficiency and durability of electrolyzers is paramount. This involves enhancing catalyst materials, optimizing cell design, and streamlining manufacturing processes. Additionally, scaling up production to meet growing demand can lead to economies of scale, driving down unit costs.

Integrating abundant and low-cost renewable energy sources, such as solar and wind power, into hydrogen production processes is essential for cost-effectiveness. By utilizing excess renewable electricity during periods of high generation or low demand, electrolysis plants can operate at optimal capacity, maximizing efficiency and minimizing costs per unit of hydrogen produced.

Large-scale deployment of hydrogen production facilities can significantly reduce capital and operational costs. Investments in infrastructure, such as electrolysis plants and renewable energy farms, enable higher production volumes, spreading fixed costs across a larger output. Additionally, collaborative initiatives and shared infrastructure among multiple stakeholders can further drive down costs through resource pooling and shared operational expenses.

Continued investment in R&D is essential for driving innovation and cost reduction in hydrogen production technologies. Research efforts focus on developing novel materials, catalysts, and process optimization techniques to enhance the efficiency, durability, and affordability of electrolysis systems. Furthermore, exploring advanced concepts like high-temperature electrolysis (e.g., solid oxide electrolyzers) offers the potential for higher efficiency and lower operating costs compared to conventional low-temperature electrolysis methods. Governments can play a crucial role in fostering the development of green hydrogen by providing supportive policies and financial incentives. Subsidies, grants, tax credits, and loan guarantees can stimulate private sector investments in hydrogen infrastructure and R&D projects. Additionally, regulatory frameworks that promote long-term market stability and facilitate public–private partnerships can create a conducive environment for cost-effective hydrogen production.

1.1.5 THE ENERGY VECTOR OF TOMORROW

Hydrogen has numerous applications, especially when produced via clean and sustainable pathways like water electrolysis powered by renewable and nuclear energy sources. This process, depicted in Figure 1.6, utilizes electric energy from renewable sources such as wind, solar, and hydro, which are naturally replenished on a human timescale. Solar panels produce electric energy directly, while wind and hydro plants use turbines coupled to rotating generators. Nuclear plants generate electric energy through steam-driven turbines, though nuclear fuel produces waste requiring treatment and storage. The electric energy is then converted to Direct Current (DC) for electrolyzers, which transform it into hydrogen, requiring only water. Hydrogen's versatility allows it to power fuel cells for transportation, be used directly in pharmaceutical industries, be injected into the gas grid, and store energy to manage peak production from renewable sources.

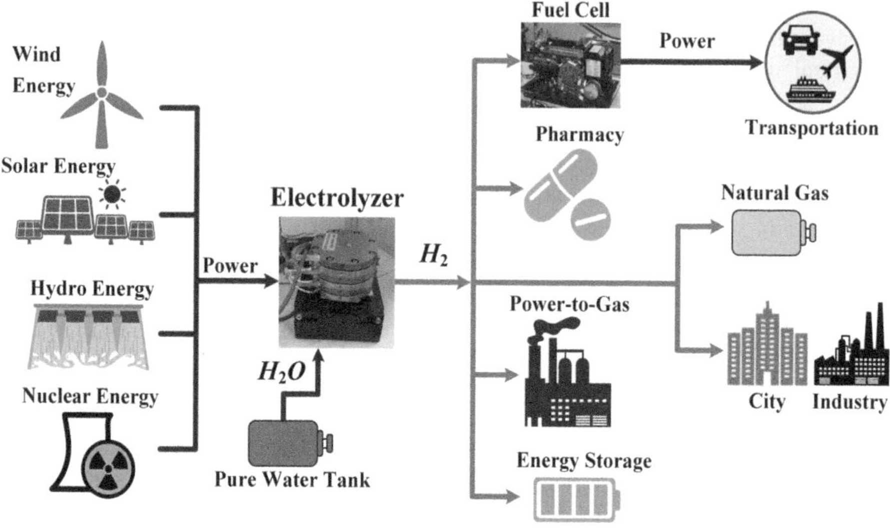

FIGURE 1.6   Overview of hydrogen applications considering renewable and nuclear power sources.

(Adapted from ref. [8].)

Figure 1.6 also highlights the carbon-free nature of hydrogen applications, emphasizing that the electrolyzer process only outputs oxygen and water, with no direct CO_2 emissions during production or utilization. This makes hydrogen an ideal energy vector for the energy transition, supporting a move toward cleaner energy systems. Hydrogen's role in storing energy and providing a carbon-free fuel alternative showcases its potential in achieving sustainable energy goals, offering a reliable and eco-friendly solution to meet various energy demands without contributing to carbon emissions.

The energy future is shaping toward hydrogen, recognized as the energy vector of tomorrow (Figure 1.7). Abundant on Earth in the form of water, the hydrogen atom stands out for its extremely energetic molecule, offering 120 MJ/kg, which is 2.5 times more than natural gas. Compared to coal, which typically yields energy content ranging between 7.8 and 8.7 kWh/kg, hydrogen demonstrates a remarkable performance approximately five times higher per unit of mass. This superior energy density renders hydrogen an exceedingly potent fuel source. Moreover, hydrogen distinguishes itself by its nonpolluting and nontoxic attributes, along with its combustion yielding only water vapor, rendering it an environmentally benign option for energy production. Additionally, hydrogen's inherent lightness, being the lightest gas, contributes to its safety profile, as it facilitates rapid diffusion in the air. In contrast to natural gas, hydrogen presents a lower risk level, particularly outside of confined environments. Its transport and storage, whether under pressure or in liquid form, are both technically feasible, further enhancing its viability as an energy carrier.

Furthermore, hydrogen exhibits particular promise as a fuel for fuel cells, representing a highly efficient means of converting chemical energy into electricity.

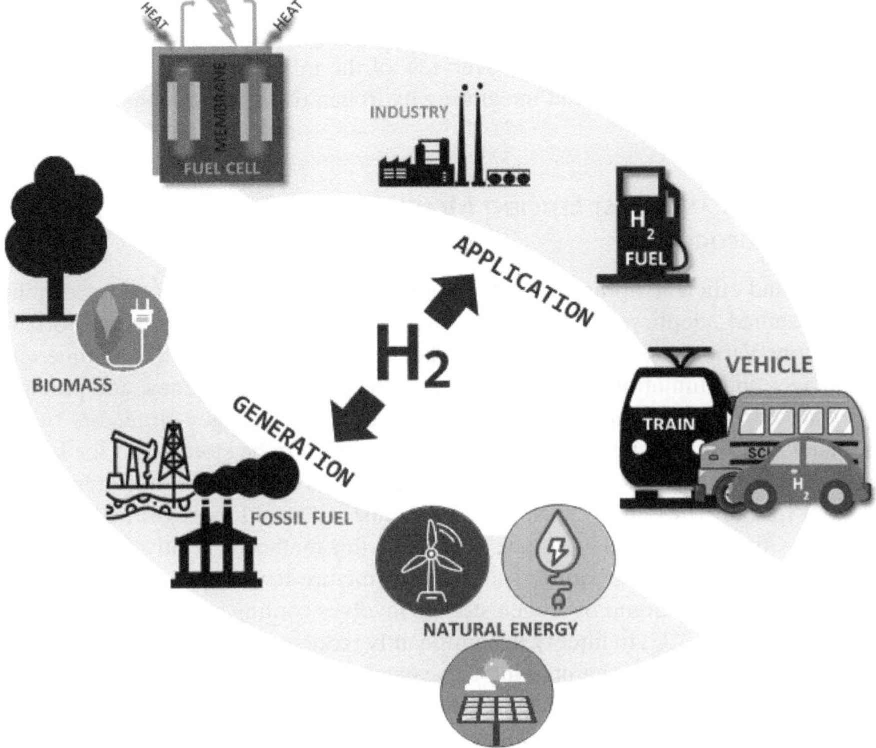

FIGURE 1.7 Hydrogen as a source of electricity.

Despite these advantages, the utilization of hydrogen in vehicles presents notable challenges, particularly in the domain of storage. First, establishing a distribution network necessitates the standardization of storage methods to ensure compatibility and efficiency across infrastructure. Second, ensuring the safety, reliability, and user-friendliness of hydrogen-powered vehicles is imperative to foster widespread acceptance and adoption. Lastly, addressing the complexities and inefficiencies associated with hydrogen storage remains a critical concern, especially in achieving long-term autonomy comparable to conventional vehicles. While techno-economic objectives for onboard hydrogen storage primarily prioritize vehicle-centric factors such as available space, weight, and costs, it is crucial to consider the capabilities and limitations of available technologies to ensure the successful integration of hydrogen as a viable energy source in transportation.

1.2   STORAGE, TRANSPORT, AND INFRASTRUCTURE

The safe and efficient storage and transport of hydrogen are essential for its adoption in various industries, including transportation and energy. This section explores effective methods for storing large quantities of hydrogen, such as compressed gas, liquid hydrogen, and solid materials. It also addresses challenges and solutions for transporting hydrogen over long distances via pipelines, tanker trucks, rail, maritime, and air transport. By examining technical, logistical, and economic aspects, this text provides a comprehensive overview of the infrastructure needed for large-scale hydrogen distribution and integrating hydrogen refueling stations into existing energy networks.

1.2.1 Safest and Most Efficient Means of Storing Large Quantities of Hydrogen

The safe and efficient storage of large quantities of hydrogen is pivotal for enabling its widespread adoption across diverse industrial sectors, including transportation and energy. Various storage methods have been developed, each with its unique set of advantages and limitations in terms of cost, safety, energy efficiency, and practicality. The selection of an appropriate storage method often hinges on factors such as the required storage duration, the volume of hydrogen to be stored, and the specific application requirements. In compressed gas storage, hydrogen is stored at high pressures in specially designed tanks. The tanks must withstand the significant pressure exerted by the compressed hydrogen gas. While this method is relatively straightforward and widely used, it requires robust infrastructure for compression and decompression processes. Liquid hydrogen storage involves cooling hydrogen to cryogenic temperatures (-253°C) to liquefy it, significantly reducing its volume for storage and transportation. However, maintaining these extremely low temperatures necessitates well-insulated storage tanks and efficient refrigeration systems. Hydrogen storage in solid materials involves the adsorption or absorption of hydrogen molecules onto or into porous materials such as metal hydrides, metal-organic frameworks (MOFs), or activated carbon. These materials offer high storage densities at moderate pressures and temperatures, making them promising candidates for onboard storage in fuel cell vehicles and portable applications.

Metal hydride salts have the ability to reversibly bind hydrogen molecules, forming stable compounds that can store and release hydrogen at ambient temperatures and pressures. This method offers advantages in terms of safety and ease of handling, as the stored hydrogen is chemically bound within the solid matrix. Underground storage involves storing hydrogen in geological formations such as salt caverns, depleted gas reservoirs, or aquifers. This method offers the advantage of large-scale storage capacity and can help to balance fluctuations in hydrogen supply and demand.

1.2.2 Methods of Safely and Economically Transporting Hydrogen over Long Distances

Transporting hydrogen over long distances presents significant technical and logistical challenges, necessitating careful consideration of various factors to ensure safe and cost-effective delivery. It is crucial to express the scientific intricacies of different techniques for long-distance hydrogen transport. Pipelines offer a highly efficient and established means of transporting large volumes of hydrogen over extensive distances. However, the existing pipeline infrastructure primarily caters to natural gas, requiring modifications to accommodate hydrogen transport. Challenges include ensuring pipeline materials' compatibility with hydrogen, as hydrogen's small molecular size can lead to embrittlement and leakage issues, necessitating the use of specialized pipeline materials and coatings. Moreover, addressing concerns related to hydrogen embrittlement in pipeline components is crucial for ensuring long-term integrity and safety.

Tanker trucks are commonly used for short- to medium-distance hydrogen transport, but their viability for long-distance haulage is limited by factors such as fuel costs, payload capacity constraints, and associated logistical complexities. Addressing these limitations may involve advancements in hydrogen storage technologies to increase onboard storage capacity and enhance energy density, thereby improving the economic feasibility of tanker truck transport for long-haul journeys. Furthermore, rail transport presents an intriguing option for long-distance hydrogen transport, leveraging existing railway infrastructures to transport hydrogen in specialized tanker cars. However, retrofitting railway systems to accommodate hydrogen transport requires substantial investments in infrastructure and rolling stock modifications. Furthermore, ensuring the compatibility of rail infrastructure materials with hydrogen and addressing safety concerns associated with hydrogen handling and storage are essential considerations.

Maritime transport offers a promising solution for regions lacking viable overland transport options. Specialized ships equipped with cryogenic tanks can transport liquid hydrogen over long distances. However, challenges include developing port infrastructure capable of handling hydrogen cargo and mitigating safety risks associated with handling cryogenic hydrogen, such as the potential for embrittlement and ignition hazards. Air transport of hydrogen is theoretically feasible for very long-distance transport or to serve remote regions inaccessible by other means. However, practical implementation faces substantial challenges, including high operational costs, limited payload capacity, and safety concerns related to hydrogen's flammability and volatility. Overcoming these challenges requires advancements in hydrogen storage and containment technologies, as well as stringent safety protocols for handling hydrogen during loading, unloading, and transit.

Each of these transport methods offers distinct advantages and challenges, requiring comprehensive risk assessments and strategic planning to ensure safe, efficient, and economically viable long-distance hydrogen transport. Collaboration among industry stakeholders, policymakers, and researchers is essential to drive innovation and address the technical, regulatory, and economic barriers hindering the widespread adoption of hydrogen as a viable energy carrier for long-distance transport applications.

1.2.3 INFRASTRUCTURE REQUIRED FOR LARGE-SCALE HYDROGEN DISTRIBUTION

The establishment of a comprehensive infrastructure for the large-scale distribution of hydrogen necessitates a systematic approach integrating various technical, logistical, and regulatory considerations. This infrastructure framework is vital to ensure the seamless production, transportation, storage, and distribution of hydrogen across different sectors. These facilities serve as the foundational nodes in the hydrogen supply chain, utilizing diverse production methods such as water electrolysis, natural gas reforming, biomass conversion, and other advanced technologies. Each production method has its unique energy requirements, environmental implications, and efficiency profiles, necessitating careful evaluation to optimize resource utilization and minimize carbon footprint. Efficient transport networks are indispensable for conveying hydrogen from production sites to distribution hubs and end users. While pipelines offer a cost-effective and environmentally friendly option for long-distance transport, the design, construction, and maintenance of hydrogen pipelines require meticulous attention to material selection, corrosion protection, and safety standards.

Distribution stations represent the interface between hydrogen producers and end users, providing essential refueling infrastructure for hydrogen-powered vehicles and industrial applications. These stations must accommodate varying demand patterns, optimize hydrogen storage and dispensing technologies, and adhere to stringent safety protocols to mitigate risks associated with hydrogen handling and storage. Hydrogen storage facilities are essential for buffering supply–demand mismatches and ensuring continuous availability of hydrogen to meet fluctuating demand. High-pressure gas storage tanks and cryogenic liquid hydrogen tanks are the primary storage technologies employed, each with its advantages and limitations in terms of storage capacity, energy efficiency, and safety considerations. End users, particularly hydrogen-powered vehicles, rely on a robust network of refueling stations for convenient access to hydrogen fuel. These stations must be strategically located, efficiently operated, and equipped with state-of-the-art dispensing equipment to ensure rapid refueling times and optimal customer experience. Standardization and regulatory frameworks are critical for ensuring interoperability, safety, and quality assurance throughout the hydrogen supply chain. Establishing harmonized standards for hydrogen production, transportation, storage, and distribution fosters market confidence, streamlines technology deployment, and facilitates cross-border trade in hydrogen-related products and services.

1.2.4 INTEGRATION OF HYDROGEN REFUELING STATIONS INTO EXISTING ENERGY NETWORKS

The integration of hydrogen refueling stations into existing energy networks demands a scientifically rigorous approach across various domains. This multifaceted process requires comprehensive planning and coordination, integrating technical, logistical, and regulatory considerations. Understanding the spatial distribution of hydrogen vehicle demand entails advanced geographical information systems techniques to assess transportation patterns, industrial concentrations, and urban demographics for optimal site selection. Ensuring seamless integration with the electrical grid involves assessing grid capacity, voltage stability, and load-balancing strategies to accommodate additional electricity demand. Establishing a reliable hydrogen supply chain involves intricate assessments of production technologies, feedstock availability, and distribution logistics, utilizing techno-economic modeling and lifecycle assessments to optimize configurations. Compliance with safety standards and regulatory requirements necessitates hazard analysis, consequence modeling, and safety barrier design to mitigate potential hazards associated with hydrogen handling and storage. Educating stakeholders on hydrogen safety protocols and raising public awareness require scientifically sound training programs and evidence-based communication strategies.

Implementing robust maintenance protocols involves predictive maintenance techniques, condition monitoring systems, and reliability engineering principles for proactive fault detection and equipment optimization. Phased deployment strategies necessitate comprehensive market analysis, technology assessment, and strategic planning, facilitated by scenario planning and adaptive management frameworks for informed decision-making and iterative refinement. Through this scientifically rigorous approach, the integration of hydrogen refueling stations can effectively advance, paving the way for a sustainable hydrogen economy.

1.3 UTILIZATION AND IMPACTS OF HYDROGEN

Each stage of the hydrogen life cycle, including production, distribution, utilization, and waste management, plays a crucial role in determining its overall sustainability. The environmental footprint of hydrogen production varies significantly based on the methods employed, with renewable energy-based electrolysis offering the most environmentally friendly option. Efficient distribution and storage systems are essential to minimize energy losses and emissions. The utilization of hydrogen in fuel cells and other applications can significantly reduce GHG emissions. Proper end-of-life management, including recycling and reuse, further enhances hydrogen's sustainability. This holistic view underscores the importance of life cycle thinking in optimizing hydrogen's role as a sustainable energy carrier and guiding policy and practice toward a greener future.

1.3.1 HYDROGEN LIFE CYCLE

The life cycle of hydrogen encompasses several stages, from its production to its end-of-life, each stage carrying environmental, economic, and social implications (Figure 1.8). The sustainability of hydrogen's life cycle will depend on choices made regarding its production, distribution, utilization, and waste management throughout its supply chain. At the production stage, the method employed to generate hydrogen greatly influences its environmental impact. For instance, hydrogen produced through electrolysis using renewable energy sources such as solar or wind power has a minimal carbon footprint compared to hydrogen derived from fossil fuels through processes like SMR, which releases GHGs.

The distribution phase involves the transportation of hydrogen to various end users, which may entail energy consumption and emissions depending on the mode of transport utilized. Efficient distribution networks, along with advancements in hydrogen storage and transport technologies, can help minimize energy loss and environmental impacts during this stage. During utilization, the efficiency of hydrogen-based technologies plays a crucial role in determining overall environmental sustainability. Fuel cell systems, for example, offer higher efficiency and lower emissions compared to traditional combustion engines. However, factors such as infrastructure development, system integration, and end-user behavior also influence the environmental performance of hydrogen utilization. Furthermore, proper end-of-life management is essential to ensure minimal environmental impact and maximize resource recovery. Recycling and reuse of hydrogen-related components

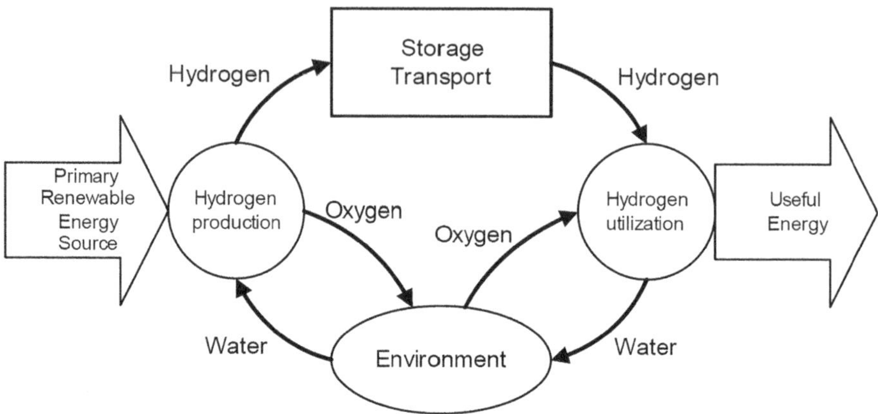

FIGURE 1.8   Diagram depicting the life cycle of hydrogen.

(Adapted from ref. [9].)

and materials can help minimize waste generation and conserve resources, contributing to the circularity of hydrogen-based systems. Accordingly, assessing and optimizing the environmental, economic, and social aspects of each stage in the life cycle of hydrogen is imperative for realizing its potential as a sustainable energy carrier. Integrating life cycle thinking into decision-making processes can help identify opportunities for improvement and guide the development of policies and practices that support a more sustainable hydrogen economy.

Hydrogen production encompasses various methods, including SMR, coal gasification, biomass gasification, and water electrolysis using renewable electricity. Each method has its own set of environmental implications, energy requirements, and resource impacts. For instance, SMR, the most common method, emits GHGs, while water electrolysis using renewable energy sources has a minimal carbon footprint. The distribution of hydrogen involves transporting it from production facilities to distribution and storage points. This can be achieved through pipelines, tanker trucks, trains, or ships for large-scale transportation. Specific infrastructure, such as hydrogen refueling stations, may be required for efficient distribution. Hydrogen storage is essential to ensure its availability during periods of high demand or when production sources are intermittent. Hydrogen can be stored in various forms, including gaseous or liquid state, or absorbed in solid materials like metal hydrides. The choice of storage method depends on factors such as cost, safety, and energy efficiency.

In terms of utilization, hydrogen finds applications across diverse sectors. It serves as a clean fuel for fuel cell vehicles, a feedstock for industrial processes like NH_3 or steel production, and a power source in fuel cells for electricity and heat generation. The environmental and economic impacts of hydrogen utilization vary depending on the sector and the specific technologies employed.

Finally, the end-of-life stage involves recycling, processing, or disposal of hydrogen. Recycling hydrogen can recover its energy or components for future use, contributing to resource conservation. Proper disposal methods may include processing to remove contaminants or impurities before safe disposal. This comprehensive understanding of the hydrogen life cycle is crucial for assessing its overall sustainability and identifying areas for improvement in its production, distribution, utilization, and end-of-life management.

1.3.2 Environmental Impacts

Hydrogen offers considerable potential as a clean energy source, but its widespread adoption will depend on resolving the challenges associated with its production, distribution, and storage. For this reason, presenting the overall impact of hydrogen usage is not straightforward. A successful transition to a hydrogen economy will require significant investments in R&D, as well as favorable policies to encourage its use while minimizing its environmental impact. Indeed, the use of hydrogen as an energy source can have a significant impact on the environment, both positive and negative. This environmental impact, from the beginning to the end of its life cycle, is influenced by various factors. The production method plays a crucial role. Benefits, such as reducing GHG emissions, can be achieved through hydrogen use in fuel cells, producing only water as a by-product. However, challenges persist, particularly regarding current hydrogen production, which often relies on fossil fuels, thus negating its environmental benefits. Nonetheless, progress is being made in developing cleaner production methods, such as electrolysis of water using renewable energy sources. The future of hydrogen as a clean energy source will depend on our ability to overcome these challenges and maximize its environmental benefits while minimizing its drawbacks.

1.3.2.1 Greenhouse Gas Emissions

The benefits of hydrogen lie in its ability to produce electricity cleanly when used in fuel cells. By reacting with oxygen from the air, it generates water as the only by-product, eliminating GHG emissions at the source, making hydrogen a clean energy source. However, its disadvantages are associated with its current production, which often relies on processes using fossil fuels such as natural gas or coal. These methods result in CO_2 emissions, thus negating the environmental benefits of hydrogen. Nevertheless, significant progress is being made in developing cleaner hydrogen production methods, notably through water electrolysis using renewable energy.

1.3.2.2 Production Impact

The advantages of hydrogen production from renewable resources such as solar, wind, or hydroelectric energy are significant, as this can greatly reduce CO_2 emissions associated with its production. Additionally, production processes can be improved to minimize energy consumption and pollutant emissions. However, there are also drawbacks to consider. For instance, large-scale hydrogen production often requires energy conversion processes that can lead to significant energy losses. This can reduce the overall system efficiency and increase the environmental footprint, despite the initial benefits. Regarding the environmental impacts of hydrogen production, several key factors can come into play. Hydrogen can be produced in various ways, each with a distinct environmental impact. The main methods include water

TABLE 1.1

Environmental Impact of Using Hydrogen

Hydrogen Production Method	CO_2 Emissions (kg CO_2/kg H_2)	Energy Consumption (MWh/ton H_2)	Water Usage (liters/kg H_2)
Steam Methane Reforming (SMR)	9–12	3–5	6–9
Coal Gasification	19–20	6–8	15–20
Electrolysis (renewable)	Near-zero	180	9
Electrolysis (nonrenewable)	Up to 10	180	9
Biomass Gasification	3–4	4–6	10–15
Solar Water-Splitting	Near-zero	50–70	9

electrolysis (using electricity from renewable, nuclear, or fossil sources) and natural gas reforming. The environmental impact will be quantified in terms of GHG emissions (mainly CO_2), energy consumption, and water usage (Table 1.1).

1.3.2.3   Distribution and Storage

Hydrogen offers significant advantages, notably its ability to be stored and transported, providing flexibility in its use and enabling its integration into various sectors such as transportation and electricity production. However, drawbacks must also be considered. For instance, the infrastructure required for hydrogen distribution and storage can pose technical and financial challenges. Additionally, there may be losses of hydrogen during storage and transport, necessitating effective solutions to minimize these losses and ensure optimal use of this resource. The environmental impact of hydrogen distribution and storage depends on several factors, such as the method of hydrogen production, the storage and transportation technologies used, and the energy source powering these processes. Table 1.2 quantitatively illustrates some of

TABLE 1.2

Environmental Impact of Distribution and Storage of Hydrogen

Distribution/ Storage Method	CO_2 Emissions (kg CO_2 per kg H_2)	Energy Consumption (kWh per kg H_2)	Other Environmental Impacts
Pipeline Transport	0.1–0.2 (depends on distance)	1–3 (depends on distance)	Environmental impact related to pipeline construction
High-Pressure Storage (cylinders)	Depends on the hydrogen source	5–10 (compression)	Use of materials for cylinders and risk of leaks
Cryogenic Storage (liquid)	Depends on the hydrogen source	10–15 (liquefaction)	Energy consumption for liquefaction and use of insulating materials
Metal Hydrides	Depends on the hydrogen source	5–15 (depends on material)	Extraction and processing of metals and risk of leaks

the environmental impacts associated with different methods of hydrogen distribution and storage.

It is crucial to note that the lowest environmental impact is generally associated with the use of hydrogen produced by water electrolysis powered by renewable energy sources, due to low CO_2 emissions and reduced consumption of fossil fuels. However, the energy efficiency and environmental impact of distribution and storage also depend heavily on the specific technologies and methods employed.

1.3.2.4 Impact on Natural Resources

Hydrogen offers notable advantages, including its ability to be produced from abundant sources such as water, thus reducing dependence on limited fossil resources. However, some drawbacks must be taken into account. For example, certain hydrogen production processes, like water electrolysis, require precious metals such as platinum, which can impact the availability and cost of these materials, potentially limiting the economic and environmental viability of these production methods.

1.3.2.5 Indirect Effects

Hydrogen offers significant potential benefits, particularly as a clean energy source that can help reduce air pollution and improve air quality, leading to direct advantages for public health. However, drawbacks must also be considered. For instance, like any emerging technology, there are uncertainties regarding the indirect environmental effects, such as ecological impacts associated with hydrogen production, storage, and usage, as well as the disposal of equipment necessary for these processes. These uncertainties underscore the need for a comprehensive assessment of the environmental and social aspects of hydrogen to ensure its sustainable and responsible use.

1.3.3 MINIMIZING CO_2 EMISSIONS

Minimizing CO_2 emissions throughout the lifecycle of hydrogen, particularly when derived from nonrenewable sources, is imperative for effective climate change mitigation. This objective requires the implementation of scientifically informed strategies aimed at mitigating carbon emissions associated with various stages of hydrogen production, distribution, and utilization. CCS stands out as a pivotal strategy, especially when hydrogen is derived from nonrenewable sources like CH_4 reforming. CCS involves capturing CO_2 emissions generated during hydrogen production and subsequently storing them in geologic formations deep underground. This method effectively prevents the release of CO_2 into the atmosphere, thus reducing the net emissions associated with hydrogen production processes.

Utilizing low-carbon energy sources, such as renewable energy (e.g., solar, wind, hydroelectric), to power hydrogen production processes offers significant potential for reducing CO_2 emissions. By leveraging renewable energy for electrolysis, hydrogen production can be decoupled from fossil fuel consumption, resulting in minimal GHG emissions during the production phase. Enhancing the energy efficiency of hydrogen production, storage, and transportation processes is critical for reducing overall CO_2 emissions. Advanced technologies, process optimization, and energy-efficient equipment can minimize energy losses and lower the carbon footprint associated with hydrogen production and distribution infrastructures. Prioritizing the production of decarbonized hydrogen through methods like electrolysis powered by renewable electricity ensures minimal CO_2 emissions during production. Decarbonized hydrogen serves as a sustainable alternative to fossil fuels in sectors where direct electrification is challenging, contributing to overall emissions reduction efforts. Continued investment in R&D is essential for advancing cleaner and more efficient hydrogen production technologies. Scientific advancements in catalysts, electrolysis processes, and carbon capture technologies hold the potential to significantly reduce CO_2 emissions across the hydrogen value chain. Establishing robust certification and traceability mechanisms ensures adherence to stringent CO_2 emission standards throughout the hydrogen lifecycle. By verifying the carbon intensity of hydrogen production processes and maintaining transparency in supply chains, stakeholders can make informed decisions and support sustainable hydrogen initiatives. By implementing these scientifically grounded approaches, it becomes feasible to effectively mitigate CO_2 emissions throughout the entire lifecycle of hydrogen, thereby supporting global efforts to combat climate change.

1.3.4 The Role of Hydrogen in a Circular Economy

Hydrogen holds significant potential to serve as a cornerstone of a circular economy, offering versatility and sustainability across various sectors. Illustrated in Figure 1.9, its pivotal position in this economic model underscores its multifaceted role.

One notable application lies in renewable energy storage, where surplus renewable electricity can be harnessed to produce hydrogen during periods of low demand or overproduction. This hydrogen serves as an efficient energy storage medium, capable of being stored and later utilized to generate electricity during peak demand or when renewable sources are less available. In the transportation sector, hydrogen plays a vital role in facilitating clean mobility. As a fuel for fuel cell vehicles, hydrogen provides a clean alternative to fossil fuels, with vehicles emitting only water vapor during operation. This contribution aids in mitigating GHG emissions and reducing air pollution, thereby advancing environmental sustainability.

Furthermore, hydrogen's versatility extends into various industrial sectors such as steel production, chemistry, and refining. Here, hydrogen can be utilized as a raw material or a reducing agent, substituting for fossil fuels. The utilization of decarbonized hydrogen in these processes can significantly diminish CO_2 emissions associated with industrial activities, aligning with the circular economy's objectives of reducing environmental impact.

Additionally, hydrogen plays a pivotal role in decentralizing the energy network. It facilitates energy storage and transport in areas where electrical infrastructure is limited or expensive to establish, thereby promoting the deployment of renewable energy sources in remote or sparsely populated regions. This decentralization aligns with the principles of a circular economy by promoting local energy production and consumption. Moreover, hydrogen contributes to the waste economy by enabling the production of hydrogen from organic waste or biomass through processes like gasification or anaerobic fermentation. This approach not only aids in organic waste

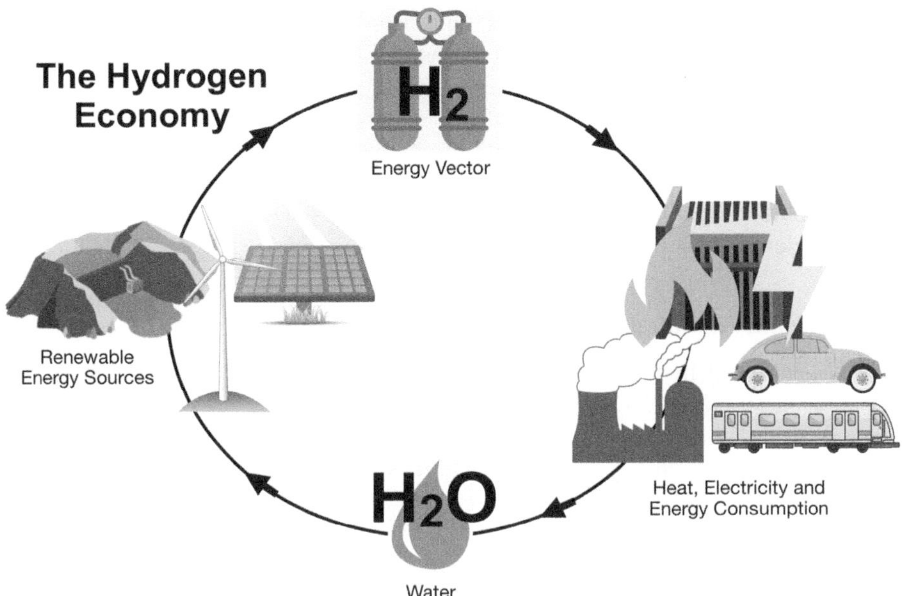

FIGURE 1.9 Hydrogen in circular economy.

(Adapted from ref. [10].)

management but also yields a clean energy source, further enhancing resource circularity and promoting sustainability.

By integrating hydrogen into various aspects of the circular economy, it becomes feasible to reduce dependence on fossil fuels, promote efficient resource utilization, and contribute to the transition toward a more sustainable and resilient economy. This integration enables the utilization of renewable energy sources for hydrogen production, effectively storing excess energy and enabling its utilization when needed. In transportation, hydrogen offers a clean alternative to traditional fossil fuels, thereby reducing GHG emissions and enhancing air quality. In industrial applications, the utilization of hydrogen as a raw material or reducing agent aids in decarbonizing processes, mitigating environmental impact. Additionally, decentralized energy networks powered by hydrogen offer increased resilience and flexibility in energy supply, especially in remote or underserved areas. Overall, by harnessing hydrogen across multiple sectors within the circular economy framework, we can foster a more sustainable and resilient future while concurrently reducing reliance on finite fossil fuel resources.

1.4 SAFETY, AWARENESS, AND TECHNOLOGICAL DEVELOPMENT

The utilization of hydrogen as an energy source introduces inherent risks that necessitate careful management and stringent safety protocols. Hydrogen's highly flammable nature poses significant hazards, including fire and explosions, demanding specialized equipment and robust fire detection and suppression systems. Environmental impacts vary with production methods, with renewable sources offering a lower carbon footprint than fossil fuel-based processes. Efficient storage and transportation of hydrogen also require advanced infrastructure and materials to ensure safety and cost-effectiveness. Comprehensive safety standards, such as ISO and National Fire Protection Association (NFPA) regulations, guide the safe handling, storage, and utilization of hydrogen. Public awareness and education, alongside continuous R&D, are critical to mitigating risks and fostering the safe, efficient integration of hydrogen technologies into the energy landscape.

1.4.1 Risks Associated with Hydrogen Use

The utilization of hydrogen as an energy source brings with it inherent risks that demand careful consideration. These risks span safety concerns, environmental impacts, and technological challenges. One of the most significant risks associated with hydrogen is its highly flammable nature, which poses hazards such as fire and explosion. To address this, stringent safety protocols must be established, governing the handling, storage, and transportation of hydrogen. This necessitates specialized equipment capable of withstanding hydrogen's unique properties, along with robust fire detection and suppression systems. The environmental implications of hydrogen production vary depending on the method employed. While hydrogen derived from renewable sources holds promise for reducing GHG emissions, processes reliant on nonrenewable sources may exacerbate environmental burdens. The scale of production often requires significant energy inputs, potentially resulting in adverse environmental impacts. Therefore, prioritizing sustainable production methodologies and energy-efficient processes is crucial.

Storage infrastructure for hydrogen, particularly at elevated pressures or in liquid form, requires specialized infrastructure. Addressing these challenges demands concerted research efforts and strategic investments. Advancements in the safety, efficiency, and cost-effectiveness of hydrogen storage and transportation technologies are paramount, necessitating the exploration of novel materials and enhancements to distribution infrastructure. By diligently investing in these areas, we can overcome current impediments and facilitate the widespread integration of hydrogen as a pivotal energy vector. This requires a holistic approach encompassing regulatory frameworks, stringent safety standards, and sustained R&D initiatives.

1.4.2 Safety Standards

Effectively handling, storing, and utilizing hydrogen demands strict adherence to comprehensive safety standards aimed at mitigating potential hazards. Hydrogen, known for its highly flammable nature, necessitates meticulous protocols to ensure operational safety and prevent accidents. Key safety standards are established to regulate various aspects of hydrogen-related activities, encompassing storage, transportation, and utilization. These standards serve as foundational guidelines, offering detailed procedures and protocols to minimize risks and safeguard against potential dangers inherent in working with hydrogen. By adhering to these standards, organizations and individuals can establish robust safety protocols, thereby fostering a secure environment conducive to the responsible use of hydrogen technologies.

1.4.2.1 ISO 22734

The ISO 22734 standard sets requirements for electrolyzers producing hydrogen through water electrolysis. It ensures safe, reliable, and efficient hydrogen production for energy purposes. Safety measures cover electrical, mechanical, thermal, and chemical risks. Reliability criteria ensure consistent and efficient hydrogen production. Interoperability standards facilitate equipment integration across manufacturers. The standard addresses electrolyzer design, construction, and materials. Operational procedures include startup, shutdown, monitoring, and maintenance. Testing methods evaluate energy efficiency and hydrogen purity. Detailed safety guidelines prevent accidents and include leak detection systems. ISO 22734 is crucial for safe hydrogen production and supports the energy transition.

1.4.2.2 NFPA 2

The NFPA 2 standard, also known as the Hydrogen Code, is a regulation developed by the NFPA of the United States. It addresses the production, storage, transfer, use, and handling of hydrogen in various contexts. The primary objective of this standard is to ensure safety when using hydrogen, a highly flammable gas, by preventing fire and explosion risks. The key points of the NFPA 2 standard include:

- Scope and application: The NFPA 2 standard applies to all facilities where hydrogen is produced, stored, transferred, or used. This includes hydrogen refueling stations, hydrogen production sites, research laboratories, and industrial facilities.
- Design and construction of facilities: The standard establishes specific requirements for the design and construction of facilities handling hydrogen to minimize fire and explosion risks. This includes specifications for ventilation systems, fire-resistant materials, and safety distances.
- Detection and suppression systems: It mandates the installation of appropriate hydrogen detection and fire suppression systems to quickly respond to hydrogen leaks or fire incidents.
- Operations and maintenance: Guidelines are provided to ensure safe operations and regular maintenance of hydrogen equipment and facilities, including staff training on safety procedures.
- Emergency measures: The standard emphasizes the importance of emergency preparedness and response involving hydrogen, including emergency response plans and coordination with local emergency services.
- Risk assessment: It highlights the need to assess risks associated with hydrogen activities and implement mitigation measures based on this assessment.

The NFPA 2 standard is regularly updated to incorporate the latest research and technologies, as well as to reflect lessons learned from new incidents and accidents. Like all NFPA standards, the goal is to promote the safety of people and property by providing evidence-based guidance and best practices in the field of hydrogen production.

1.4.2.3   ISO 19880-1

The ISO 19880-1 standard, also known as "Hydrogen fueling stations - Part 1," is a globally recognized framework developed by the International Organization for Standardization (ISO) to ensure the safe and efficient operation of hydrogen fueling stations. It encompasses a comprehensive set of guidelines and requirements for the design, construction, operation, and maintenance of these crucial infrastructures that supply hydrogen to vehicles, including fuel cell cars and buses. With a primary focus on safety, the standard emphasizes stringent measures throughout all stages, from initial planning to routine maintenance, to mitigate potential risks associated with handling hydrogen, a highly flammable gas, and to safeguard both public welfare and environmental integrity.

The key components of the ISO 19880-1 standard include stipulations for general requirements, safety protocols, performance and reliability criteria, interoperability standards, maintenance guidelines, and personnel training recommendations. These specifications are meticulously designed to ensure the consistent delivery of hydrogen fueling services while minimizing hazards and vulnerabilities. By providing a universal framework, the standard facilitates the harmonization of practices across different regions and promotes international collaboration in advancing hydrogen fueling infrastructure. Furthermore, it serves as a dynamic document that evolves in tandem with technological advancements and industry insights, reinforcing its role as a cornerstone in the global transition toward sustainable hydrogen-based transportation systems.

1.4.2.4   ISO 16111

The ISO 16111 standard, titled "Transportable storage systems - Hybrid hydrogen storage units," specifies the requirements for hydrogen storage devices that are transportable and can be used for temporary hydrogen storage in various applications. This standard covers a range of topics, including the materials used for storage, the design and construction of the devices, testing methods to verify safety and performance, as well as required markings and documentation. Specifically addressing hybrid hydrogen storage units, the ISO 16111 standard covers systems that can utilize more than one method of hydrogen storage within the same device. Hydrogen storage methods include high-pressure storage (in high-pressure tanks), liquid storage (at very low temperatures), and storage in solid materials (where hydrogen is adsorbed or chemically bound to a material). The key aspects of the ISO 16111 standard include safety requirements, detailing safety requirements for the design, manufacturing, and operation of hydrogen storage units to minimize the risks of fire, explosion, or hydrogen leakage. It also encompasses performance and testing, providing standardized testing methods to assess storage capacity, durability, and reliability. Additionally, the standard addresses marking and documentation requirements, ensuring that necessary information is provided with hydrogen storage units for proper use and maintenance. The design and manufacturing criteria are established to ensure that hydrogen storage devices are constructed to meet performance and safety requirements, and material compatibility is addressed to prevent degradation or undesirable chemical reactions.

This standard plays a crucial role in facilitating the safe and efficient use of hydrogen as an energy carrier, particularly in applications where hydrogen needs to be transported or stored temporarily. By establishing clear criteria and safety requirements, the ISO 16111 standard contributes to promoting the adoption of hydrogen in various sectors, including transportation, renewable energy, and industry.

1.4.2.5 ISO 16110

The ISO 16110 standard, titled "Hydrogen production systems," outlines requirements for the safe and efficient operation of hydrogen production systems, particularly those utilized in vehicle refueling applications. Divided into two main parts, the standard covers general requirements applicable to all hydrogen production systems and specific criteria for systems based on water electrolysis. It emphasizes safety, efficiency, reliability, and maintenance as key considerations in the design, installation, operation, and maintenance of hydrogen production systems.

The first part of the standard, ISO 16110-1, provides general requirements applicable to all hydrogen production systems. It addresses safety measures to mitigate risks associated with hydrogen production, such as fire, explosion, and hydrogen leakage. Additionally, it underscores the importance of efficiency and reliability in ensuring continuous hydrogen supply. Maintenance guidelines are also outlined to sustain the long-term functionality of hydrogen production systems.

ISO 16110-2, the second part of the standard, focuses specifically on hydrogen production systems utilizing water electrolysis. It details design specifications, installation guidelines, operational recommendations, and testing methods tailored to this method of hydrogen production. By providing clear criteria for the design, installation, operation, and maintenance of hydrogen production systems, the ISO 16110 standard aims to promote the safe and efficient deployment of hydrogen technology, particularly in applications involving fueling fuel cell vehicles.

1.4.2.6 EN 17124

The EN 17124 standard, a European norm governing locks and door fittings, sets out performance criteria and testing methodologies for these components used in various building applications. Covering both mechanical and electronic locks, it ensures they meet stringent requirements for burglary resistance, durability, corrosion resistance, and operational security. By outlining standardized test procedures and classification systems, the standard helps users assess the performance levels of locks and locking devices, facilitating informed decisions about product selection based on specific needs and requirements. Moreover, products compliant with EN 17124 can display the CE marking, signifying adherence to European standards and regulations, particularly vital for products utilized in public and commercial buildings across the European Union (EU). In essence, the standard plays a pivotal role in upholding the quality and safety of locks and locking mechanisms, fostering confidence in their reliability and effectiveness within building infrastructure.

1.4.2.7 NFPA 55

The NFPA 55 standard, managed by the NFPA, regulates the handling, storage, usage, manipulation, and control of compressed and liquefied gases in portable containers. Covering a broad spectrum of gases, including flammable, nonflammable, toxic, and corrosive varieties, NFPA 55 is instrumental in mitigating risks associated with these substances, such as fire hazards, explosion potential, toxicity, and hazardous chemical reactions. The standard classifies gases based on their properties and prescribes safety measures accordingly, addressing storage quantities, separation requirements, storage facility design, and ventilation needs, among other aspects. Additionally, NFPA 55 mandates training for personnel, installation of detection systems, and implementation of emergency plans to effectively respond to incidents involving compressed or liquefied gases, ensuring comprehensive safety protocols across industries.

NFPA 55 is a pivotal framework for businesses and professionals engaged in working with compressed and liquefied gases, offering clear guidelines and regulations to safeguard operations and protect workers and the wider community. Regular updates to the standard reflect advancements in technology, evolving scientific insights, and emerging safety practices, underscoring its commitment to maintaining relevance and efficacy in the face of changing industrial landscapes. Compliance with NFPA 55 is essential for ensuring the safe handling and usage of compressed and liquefied gases, thereby upholding operational integrity and minimizing risks associated with these potentially hazardous materials.

1.4.2.8 NFPA 52

This standard provides requirements for the design, construction, maintenance, and inspection of hydrogen fueling systems for hydrogen vehicles. The NFPA 52 standard, developed and published by NFPA, specifically addresses vehicular natural gas fuel systems. It aims to establish safety practices for storing and handling natural gas (both compressed and liquefied) used as fuel for vehicles, covering various aspects such as design, construction, installation, operation, inspection, and maintenance of vehicular natural gas fuel systems and fueling facilities. The NFPA 52 standard is designed to ensure a high level of safety in the use of natural gas as a vehicular fuel, reducing risks of fire, explosion, and exposure to toxic gases. It undergoes regular revisions and updates to incorporate the latest technological advancements and industry best practices, providing comprehensive guidelines for ensuring the safe operation and maintenance of natural gas fuel systems and fueling stations.

1.4.2.9 EN ISO 13985

The European standard EN ISO 13985 establishes safety requirements for hydrogen refueling stations catering to hydrogen vehicles fueled by compressed gas. These standards, regularly updated to incorporate the latest technological advancements and safety practices, are vital for ensuring the safety of individuals, facilities, and the environment within the hydrogen industry. They cover various aspects such as structural safety, material compatibility, refueling procedures, continuous monitoring, fire prevention, leak detection, emergency response plans, maintenance programs, personnel training, and environmental considerations. Compliance with standards like EN ISO 13985 is crucial for the safe and effective deployment of hydrogen infrastructure, facilitating the transition to hydrogen vehicles as a clean alternative to fossil fuels.

The standard plays a pivotal role in ensuring the safe design, construction, and operation of hydrogen refueling stations, which are essential components of the infrastructure supporting hydrogen-powered vehicles, including cars, buses, and trucks. The key elements covered include structural safety guidelines, material selection for compatibility with hydrogen, safe refueling protocols, continuous monitoring systems, fire prevention measures, leak detection installations, emergency response planning, regular maintenance programs, personnel training, and environmental impact considerations. These standards, reflecting international consensus on best practices, are regularly updated to incorporate the latest technological advancements and operational experiences, further enhancing safety and efficiency in the hydrogen industry.

1.4.3 REDUCING RISKS ASSOCIATED WITH HYDROGEN HANDLING

Handling hydrogen requires meticulous attention due to its highly flammable nature and potential safety hazards. Comprehensive training programs should be implemented to educate individuals involved in handling hydrogen on safety protocols, emergency procedures, and proper storage techniques. Training should cover the characteristics of hydrogen, its potential hazards, and methods for safe handling. Adequate ventilation systems must be in place to prevent the buildup of hydrogen gas, which is highly flammable in confined spaces. Proper ventilation helps dissipate any leaked hydrogen and reduces the risk of ignition. Specialized containers and equipment designed specifically for handling hydrogen safely are essential. These should be constructed from materials resistant to hydrogen embrittlement and corrosion. Regular inspection and maintenance of containers and equipment are necessary to ensure their integrity.

Conducting thorough risk assessments is critical to identify potential hazards associated with hydrogen handling activities. Implementing robust risk management strategies involves identifying potential risks, assessing their likelihood and severity, and implementing controls to mitigate them effectively. Minimizing the presence of ignition sources in hydrogen handling areas is crucial. This includes eliminating open flames, electrical sparks, and hot surfaces that could ignite hydrogen gas. Implementing strict protocols for hot work permits and conducting hazard assessments before any activity that could generate sparks or heat are essential precautions. Employing reliable leak detection systems, such as hydrogen sensors, can help quickly identify and localize any leaks. Establishing clear procedures for responding to hydrogen leaks, including immediate evacuation and isolation measures, is imperative to prevent accidents. Hydrogen storage facilities should adhere to stringent safety standards and regulations. Properly designed and ventilated storage areas, equipped with suitable containment systems, minimize the risk of hydrogen accumulation and potential ignition sources.

Fostering a culture of safety through effective communication and awareness programs is vital. Ensuring that all personnel are knowledgeable about hydrogen safety protocols and actively engage in safety practices promotes a safer working environment. By implementing these scientifically informed measures, the risks associated with handling hydrogen can be effectively managed, contributing to a safer workplace environment and minimizing the potential for accidents or incidents.

1.4.4 Public Awareness and Education

Raising awareness and disseminating knowledge about hydrogen's role as an energy carrier requires a multifaceted approach that prioritizes scientific rigor. One effective strategy involves developing comprehensive campaigns that integrate scientific data, visual representations, and easily understandable content. These campaigns aim to elucidate hydrogen's properties, production processes, and potential applications, with a particular emphasis on its pivotal role in decarbonizing various sectors and mitigating GHG emissions.

Another crucial tactic is organizing specialized symposiums, workshops, and seminars aimed at diverse stakeholders, including policymakers, industry experts, researchers, and the general public. These events represent the scientific underpinnings of hydrogen technologies, featuring presentations on advancements in electrolysis, fuel cell efficiency, and hydrogen storage methodologies. By providing a platform for in-depth discussion and knowledge exchange, these gatherings play a vital role in fostering a deeper understanding of hydrogen's potential and challenges.

Furthermore, integrating hydrogen-related topics into science, technology, engineering, and mathematics (STEM) education at both secondary and tertiary levels is essential. This involves developing curricular modules with interactive learning materials and laboratory experiments that facilitate hands-on exploration of hydrogen's chemical properties, energy conversion mechanisms, and environmental implications. By engaging students early on, educators can help cultivate a future workforce equipped to drive innovation in hydrogen technology and contribute to sustainable development efforts.

1.4.5 Strategies for Hydrogen Technology Development in Various Sectors

To effectively utilize hydrogen across diverse sectors, various key strategies must be implemented. Adopting a robust approach to R&D is crucial. This involves investing in innovative technologies to enhance efficiency and reduce costs associated with hydrogen production, storage, distribution, and utilization. Also, developing a comprehensive infrastructure network is essential. This includes constructing hydrogen refueling stations, pipelines, and storage facilities to support the widespread adoption of hydrogen across different sectors. Establishing standardized regulations is vital to ensure the safety, quality, and interoperability of hydrogen technologies. Consistent standards for production, storage, transportation, and utilization facilitate market integration and promote confidence among stakeholders.

Providing training and raising awareness among industry professionals, policymakers, and the public is essential. Comprehensive education programs and awareness campaigns promote the understanding of hydrogen technologies and their benefits, dispelling misconceptions and garnering support. Fostering collaboration among governments, industry stakeholders, academia, and NGOs drives innovation and accelerates hydrogen deployment. Public–private partnerships leverage expertise and resources to scale up projects and demonstrate the viability of hydrogen technologies. Implementing economic incentives and supportive policies stimulates investment and adoption of hydrogen technologies. Subsidies, tax credits, and deployment targets create market demand, driving long-term growth and sustainability in the hydrogen sector.

1.5 GOVERNMENT POLICIES AND FUTURE VISION

Effective government policies are crucial for the development and integration of hydrogen technology into the global energy landscape. This section outlines the essential measures needed to promote hydrogen, such as investments in research, financial incentives, infrastructure development, and regulatory standards. It also highlights the strategic initiatives of key players like the United States, China, Japan, and Europe. Additionally, the section explores the anticipated energy landscape of 2050, emphasizing hydrogen's potential to play a pivotal role in achieving a cleaner, more sustainable, and resilient energy future, driven by renewable sources and technological advancements.

1.5.1 ESSENTIAL GOVERNMENT POLICIES FOR HYDROGEN TECHNOLOGY DEVELOPMENT

Promoting hydrogen development requires coordinated government measures. Key actions include investments in R&D, which are essential for driving innovation and reducing costs associated with hydrogen production, storage, distribution, and utilization. Governments should fund universities, research institutions, and private sector initiatives to develop cutting-edge technologies that enhance hydrogen efficiency and viability. Financial incentives, such as subsidies and tax credits, are also crucial for encouraging adoption in various sectors like transportation, industry, and energy. These incentives could include grants for purchasing hydrogen equipment, tax breaks for companies investing in hydrogen infrastructure, and subsidies for consumers choosing hydrogen-powered vehicles.

Infrastructure development is another critical component, requiring the establishment of hydrogen refueling stations and storage facilities. Governments need to invest in building a robust network of refueling stations to support hydrogen-powered vehicles, along with storage facilities that ensure a steady supply of hydrogen. Ensuring the safety, quality, and interoperability of hydrogen technologies through clear regulatory standards is also vital. These standards should cover all aspects of hydrogen use, from production to storage and transportation, ensuring that hydrogen technologies are safe and reliable.

Public–private partnerships are necessary to foster research initiatives and pilot projects, encouraging collaborations between the public and private sectors to share knowledge, resources, and expertise in developing hydrogen technologies. Lastly, awareness and education campaigns are essential to inform the public and businesses about the benefits of hydrogen as a clean and sustainable energy carrier. Educational campaigns should highlight hydrogen's potential to reduce carbon emissions and its role in transitioning to a renewable energy future.

1.5.1.1 United States

The US government's hydrogen policy, positions clean hydrogen as essential for the nation's clean energy future. The Department of Energy (DOE) has allocated $7 billion to establish seven Regional Clean Hydrogen Hubs (H2Hubs) to expedite deployment, particularly in industrial processes and heavy-duty transportation. These hubs aim to create localized hydrogen economies, supporting industries and creating jobs.

The National Clean Hydrogen Strategy and Roadmap sets ambitious targets for 2030, 2040, and 2050, emphasizing collaboration among federal agencies, industry, and academia to catalyze market penetration and technological breakthroughs. The strategy includes specific goals such as reducing the cost of hydrogen production, expanding infrastructure, and increasing hydrogen use in various sectors. Financial commitments include $9.5 billion from the Bipartisan Infrastructure Law and additional incentives from the Inflation Reduction Act. These funds aim to drive innovation, lower costs, and support infrastructure development. The strategy projects the creation of 100,000 new jobs by 2030, underscoring the economic benefits of a robust hydrogen sector.

1.5.1.2 China

China's long-term hydrogen plan (2021–2035) aims to develop a domestic hydrogen industry and master related technologies, driven by commitments to carbon peak and neutrality. The plan outlines a phased approach to enhance innovation capacity and production capabilities. By 2025, China plans to establish a comprehensive hydrogen energy system with significant production from renewable sources, potentially reducing CO_2 emissions by up to 2 million metric tons annually. The country aims to master key technologies and manufacturing processes, ensuring a stable supply of hydrogen. By 2030, China seeks to achieve a reasonable and orderly industrial layout, integrating hydrogen into its energy mix to support its carbon peak objective. By 2035, the proportion of hydrogen produced from renewable sources in terminal energy consumption is expected to significantly increase. The plan includes targets for 50,000 fuel cell electric vehicles (FCEVs) and 100,000 to 200,000 metric tons of green hydrogen annually by 2025. Financial assistance and government incentives will support these goals. This plan positions China as a global leader in hydrogen technology, bolstering its low-carbon economy and energy sector. An illustration of China's roadmap, along with pertinent policy and strategy benchmarks, is presented in Figure 1.10.

1.5.1.3 Japan

Japan's strategy, adopted in 2017, aims to transform into a "hydrogen society" by increasing energy efficiency, enhancing energy security, reducing GHG emissions, and developing hydrogen-based industrial activities. The strategy includes widespread deployment of hydrogen vehicles, construction of hydrogen stations, and development of hydrogen as an energy source for residences and mobility. The

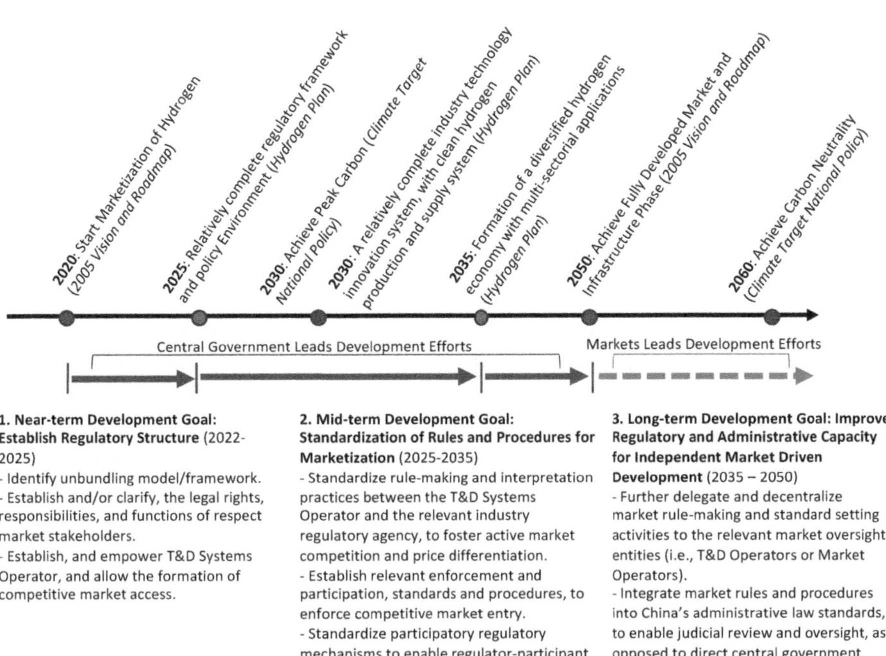

1. Near-term Development Goal: Establish Regulatory Structure (2022-2025)
- Identify unbundling model/framework.
- Establish and/or clarify, the legal rights, responsibilities, and functions of respect market stakeholders.
- Establish, and empower T&D Systems Operator, and allow the formation of competitive market access.

2. Mid-term Development Goal: Standardization of Rules and Procedures for Marketization (2025-2035)
- Standardize rule-making and interpretation practices between the T&D Systems Operator and the relevant industry regulatory agency, to foster active market competition and price differentiation.
- Establish relevant enforcement and participation, standards and procedures, to enforce competitive market entry.
- Standardize participatory regulatory mechanisms to enable regulator-participant exchange for market developments.

3. Long-term Development Goal: Improve Regulatory and Administrative Capacity for Independent Market Driven Development (2035 – 2050)
- Further delegate and decentralize market rule-making and standard setting activities to the relevant market oversight entities (i.e., T&D Operators or Market Operators).
- Integrate market rules and procedures into China's administrative law standards, to enable judicial review and oversight, as opposed to direct central government intervention.

FIGURE 1.10   Proposed regulatory pathway to China's emerging hydrogen economy. **(Adapted from ref. [11].)**

Japanese government has set ambitious goals, such as equipping 5.3 million households and registering 800,000 hydrogen vehicles by 2030. Significant funds are allocated for hydrogen development, including 70 billion yen from the Ministry of Economy, Trade, and Industry (METI) and 6.58 billion yen from the Ministry of the Environment. Japan also focuses on reducing hydrogen production costs to make this energy competitive with fossil fuels while enhancing energy security and contributing to GHG emissions reduction. International collaboration and private sector participation are key elements of this strategy, promoting hydrogen use in various economic sectors and encouraging investments in hydrogen-related technologies.

Japan's envisioned low-carbon society relies primarily on nuclear, renewable, and fossil energy with CCS as primary energy sources. Renewable energy plays a crucial role, with significant capacities in solar, wind, geothermal, hydroelectric, and biomass energy. The integration of hydrogen into the energy mix enhances Japan's energy security by reducing reliance on imported fossil fuels.

By 2050, hydrogen is expected to account for 13% of Japan's total primary energy supply, highlighting its pivotal role in achieving a low-carbon society. The strategic shift to hydrogen-based energy systems capitalizes on advanced technologies for production, storage, and utilization, reinforcing Japan's commitment to sustainable energy practices and technological innovation.

1.5.1.4 Europe

Europe's policy on hydrogen technology development aims to make hydrogen a key part of its transition to a carbon-neutral economy by 2050, as outlined in the EU Hydrogen Strategy (2020). The strategy includes three phases: 2020–2024, focusing on installing at least 6 GW of renewable hydrogen electrolyzers to produce up to one million tons of renewable hydrogen; 2025–2030, scaling up production with at least 40 GW of electrolyzers to produce up to ten million tons of renewable hydrogen; and 2030–2050, deploying hydrogen at large scale in all hard-to-decarbonize sectors. Key sectors include transport, industry, energy storage, and building heating. The European Commission proposes an initial investment of 430 billion euros by 2030 to support the production and integration of renewable hydrogen in these sectors.

In transport, the focus is on reducing CO_2 emissions from medium- and long-distance vehicles, including trucks, buses, trains, and potentially airplanes and ships. For industry, the aim is to decarbonize energy-intensive industrial processes such as steelmaking, chemicals, and refining. In the energy sector, hydrogen will be used to store surplus energy from renewable sources, providing long-term energy storage and facilitating grid balancing. In building heating, there is potential for using hydrogen to heat residential and commercial buildings through district heating networks or by blending hydrogen into the natural gas grid. To support the transition, the EU plans to establish industrial alliances like the European Clean Hydrogen Alliance, finance critical infrastructure projects through European financial instruments such as the Connecting Europe Facility (CEF) and the Innovation Fund, and create a regulatory framework that encourages investment in clean hydrogen (Figure 1.11) [13].

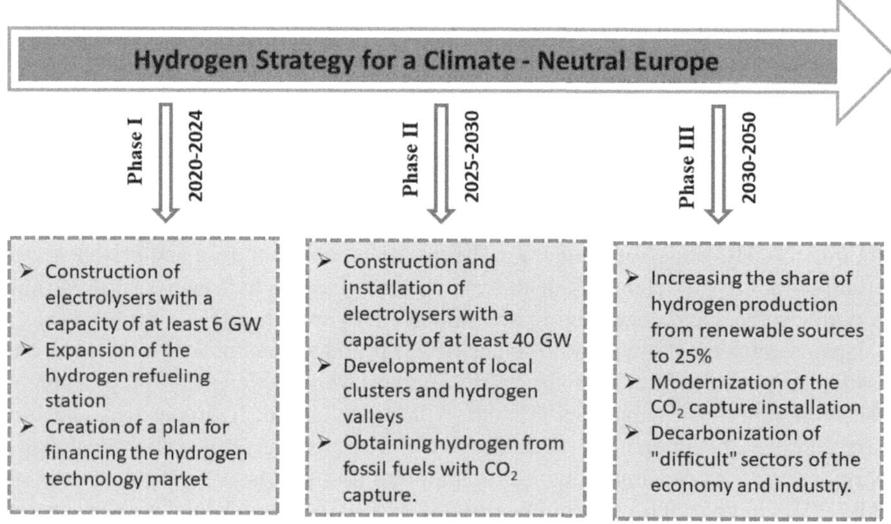

FIGURE 1.11 Phases of the EU Hydrogen Strategy for 2020–2050.

(Adapted from ref. [13].)

1.5.2 LIKELY ENERGY LANDSCAPE IN 2050 AND THE ROLE OF HYDROGEN

The energy landscape of 2050 is subject to much speculation and analysis based on current trends, political commitments to reduce GHG emissions, technological advancements, and changes in energy consumption patterns. While it is difficult to predict with certainty, several key themes emerge regarding the role of hydrogen and the overall direction of the energy landscape. The transition to an energy mix dominated by renewable energies (solar, wind, hydroelectric, etc.) is expected to accelerate, thereby reducing dependence on fossil fuels. This transition is supported by commitments made under the Paris Climate Agreement and the carbon neutrality goals set by many countries. The electrification of transportation, heating, and many other sectors, coupled with ongoing improvements in energy efficiency, will play a crucial role in reducing CO_2 emissions. Smart technologies and more flexible electrical grids will enable better integration of renewable energies.

Hydrogen, especially green hydrogen produced from renewable energy sources, is likely to play a significant role in the future energy landscape for several reasons. Hydrogen can serve as a long-term energy storage medium for renewable energy, overcoming the intermittency of sources such as solar and wind. Some industrial sectors (steel, cement) and modes of transportation (maritime, aviation, and certain heavy-duty road transport) are difficult to electrify and can benefit from hydrogen as a clean energy carrier. Hydrogen can be transported and stored using infrastructure that, in some cases, can be adapted from existing natural gas systems, offering additional flexibility in energy distribution.

The integration of energy systems is likely to be more pronounced, with close links between energy production, storage, distribution networks, and consumers. Hydrogen could play a key role in these integrated systems, facilitating the cross-sectoral use of energy. Cost reductions and technological advancements will be decisive for the adoption of hydrogen and other renewable energies. Favorable policies and regulations are needed to support investment in clean technologies and hydrogen infrastructure. The energy transition will also require public support and consideration of social and economic impacts.

Accordingly, while many uncertainties remain, it is clear that hydrogen has the potential to play an important role in a future energy landscape that is cleaner, more sustainable, and resilient to the challenges of climate change. Achieving climate neutrality in the EU within a short timeframe presents a formidable challenge, necessitating a comprehensive long-term strategy. The European Council underscores the urgency of transforming the industrial sector within 25 years, emphasizing immediate action. The hydrogen sector, pivotal in the European Green Deal, promises substantial contributions to energy sustainability. This involves enhancing renewable energy technologies, which require storage solutions for use during energy deficits in the power grid. The expansion of road, rail, and water transport, coupled with EU ambitions to reduce GHGs, is expected to increase demand for hydrogen technologies as alternative fuels. Financial investment in such technologies could lead to significant advancements in hydrogen usage or storage. However, all timelines and phases of implementation are contingent on the current state of knowledge and can vary based on unforeseen developments or innovations.

On July 8, 2020, the European Commission unveiled the "Hydrogen Strategy for a Climate-Neutral Europe," aiming to boost the green hydrogen sector's contribution to the EU's energy mix from 2% to 13–14% by 2050, enhancing decarbonization and air quality while reducing CO_2 emissions. This strategy, however, faces legal, social, and financial challenges, primarily due to the high costs of hydrogen production and the need to expand the electrolyzer production sector. Despite these challenges, the sector's growth, driven by international collaboration and investment, could spur economic growth. The strategy also outlines the significance of producing different types of hydrogen, like green, blue, and gray, each with distinct environmental impacts and production methods, impacting their future viability and environmental benefits.

The US DOE recently unveiled its green hydrogen program plan, aiming to expedite R&D projects that lay the foundation for producing, distributing, storing, and utilizing hydrogen from various sources for short-, mid-, and long-term applications. The US government is committed to decreasing the levelized cost of hydrogen production, delivery, storage, and conversion systems, while enhancing their performance and durability. The DOE's focus on green hydrogen includes overcoming technological, socioeconomic, and safety challenges that hinder the integration of hydrogen with traditional energy systems, as well as exploring hydrogen export opportunities and its broad applications. The program adopts a comprehensive approach, targeting all aspects of hydrogen including production, delivery, storage, conversion, and applications. It sets an ambitious target to produce carbon-neutral blue hydrogen at 1 dollar per kilogram by 2025. Additionally, the plan aims for green hydrogen production costs to be less than 2 dollars per kilogram for transportation and less than 1 dollar per kilogram for industrial feedstock use by 2030. The plan also emphasizes developing lower-cost hydrogen pipelines, long-term storage solutions, and the blending of hydrogen gas into existing natural gas pipelines. Moreover, it includes a target of less than 80 dollars per kilowatt for manufacturing fuel cells that utilize hydrogen, along with the installation of 4,000 hydrogen fueling stations by 2030.

This detailed and coordinated approach by the United States underscores the global commitment to integrating hydrogen into the future energy landscape, driving both economic growth and environmental sustainability.

1.5.3 HYDROGEN TECHNOLOGY IN INDUSTRY 4.0 AND THE INDUSTRY OF THE FUTURE (INDUSTRY 5.0)

Hydrogen technology is poised to become a cornerstone of Industry 4.0, the current trend of automation and data exchange in manufacturing technologies. This transformation is driven by advancements in cyber-physical systems, the Internet of Things (IoT), and cloud computing. Hydrogen technology supports this transition by offering a versatile, sustainable, and efficient energy source. Here, we explore the role of hydrogen in Industry 4.0 and the future industry landscape. Although hydrogen possesses remarkable power generation capabilities, only a small fraction is employed for this purpose. A substantial amount of commercially produced hydrogen is utilized in diverse industries, such as metalworking, oil refining and recycling, fertilizers, and chemical processing (Figure 1.12).

FIGURE 1.12 Current and future industrial applications of hydrogen.

(Adapted from ref. [15].)

1.5.3.1 Industry 4.0 Characteristics and Hydrogen's Role

Industry 4.0 emphasizes the integration of cyber-physical systems, the IoT, cloud computing, and cognitive computing. These technologies enable the creation of smart factories with interconnected machinery and advanced data analytics, optimizing production processes and improving efficiency. Hydrogen technology aligns perfectly with these principles by providing a clean, efficient, and flexible energy source.

Hydrogen fuel cells offer a dependable and eco-friendly power solution for autonomous robots, smart machines, and various industrial equipment. These fuel cells are particularly advantageous in environments where traditional power sources are either impractical or insufficient. For instance, in remote locations or in situations requiring uninterrupted power supply, hydrogen fuel cells ensure continuous operation without the drawbacks of conventional energy sources, such as noise, emissions, or the need for frequent refueling. Furthermore, hydrogen technology plays a crucial role in high-temperature industrial processes, such as those in the steel and chemical industries. These processes traditionally rely on fossil fuels to achieve the necessary high temperatures, resulting in significant carbon emissions. By substituting fossil fuels with hydrogen, these industries can not only meet their high-temperature requirements but also significantly reduce their carbon footprint. This transition to hydrogen aligns with Industry 4.0's focus on sustainability, as it promotes the use of cleaner energy sources and enhances overall energy efficiency. Hydrogen's ability to provide a clean and efficient energy source for both autonomous systems and high-temperature industrial processes illustrates its versatility and importance in modern manufacturing and industrial operations. By integrating hydrogen technology, industries can achieve greater sustainability, operational efficiency, and resilience, all of which are key objectives of Industry 4.0. This shift not only supports environmental goals but also drives innovation and competitiveness in the industrial sector.

Hydrogen serves a critical role in renewable energy storage by acting as an energy carrier that captures surplus electricity generated from renewable sources such as wind and solar. During periods of high renewable energy production, excess electricity can be used to produce hydrogen through the process of electrolysis. This hydrogen can then be stored for extended periods, offering a versatile solution to the intermittency challenges associated with renewable energy sources. When renewable energy generation is low, the stored hydrogen can be converted back into electricity using fuel cells or turbines, thereby ensuring a continuous and stable energy supply. This capability not only maximizes the utilization of renewable energy but also enhances the overall resilience of the energy system. Incorporating hydrogen storage into smart grids significantly improves grid balancing by effectively managing supply and demand fluctuations. Smart grids, equipped with advanced monitoring and control technologies, can dynamically adjust to changes in energy production and consumption. By integrating hydrogen storage, these grids can absorb excess energy during periods of low demand and release it during peak demand times. This ability to balance the grid reduces the risk of blackouts and brownouts, which are often caused by mismatches between energy supply and demand. Furthermore, it enhances the efficiency and reliability of the energy system by providing a buffer that mitigates the impact of variable renewable energy sources. The integration of hydrogen technology into energy storage and grid balancing systems represents a transformative advancement for modern energy infrastructures. It supports the broader adoption of renewable energy by addressing its inherent variability, thus contributing to a more sustainable and resilient energy landscape. By leveraging hydrogen as both an energy storage medium and a tool for grid stability, we can move toward a cleaner, more efficient, and reliable energy future.

Producing hydrogen through electrolysis powered by renewable energy allows industries to significantly reduce their carbon footprint. This process, known as green hydrogen production, involves splitting water into hydrogen and oxygen using electricity generated from renewable sources such as wind, solar, or hydroelectric power. By avoiding the use of fossil fuels in hydrogen production, industries can drastically cut down on GHG emissions. The resulting green hydrogen can be utilized in a wide range of applications, including power generation, transportation, and various industrial processes. For example, in power generation, hydrogen can be used in fuel cells to produce electricity with water as the only by-product, making it an extremely clean energy source. In transportation, hydrogen fuel cells can power vehicles, offering a zero-emission alternative to traditional ICEs. Hydrogen technology also enhances resource efficiency, contributing to a circular economy by enabling the optimal use of excess renewable energy and minimizing waste. During periods of high renewable energy production, when supply exceeds demand, the surplus energy can be used to produce hydrogen. This hydrogen can be stored and later converted back into electricity or used directly as a fuel, ensuring that no renewable energy is wasted. This capability not only improves the overall efficiency of energy use but also supports sustainable industrial practices by reducing reliance on nonrenewable resources and minimizing environmental impact. Furthermore, hydrogen technology supports resource conservation by enabling industries to adopt more sustainable practices. For instance, hydrogen can replace fossil fuels in high-temperature industrial processes, such as steel and cement production, which are traditionally associated with high carbon emissions. By using green hydrogen, these industries can significantly lower their environmental footprint and move toward more sustainable production methods. Additionally, the use of hydrogen in industrial processes can help conserve water and other resources by reducing the need for water-intensive cooling systems typically associated with fossil fuel-based power generation.

1.5.3.2 Future Industry Landscape (Industry 5.0)

Industry 5.0 builds on the foundations of Industry 4.0 by emphasizing human-centric, sustainable, and resilient manufacturing. It integrates human intelligence with advanced technologies to create a more personalized and environmentally friendly industrial landscape. Hydrogen technology facilitates decentralized energy production, enabling industries to generate and utilize hydrogen on-site. This localized production approach offers significant flexibility, allowing industries to tailor their energy solutions to specific operational needs. By producing hydrogen directly at the point of use, industries can reduce the inefficiencies associated with long-distance transportation and distribution of energy. This not only minimizes energy losses but also lowers the overall carbon footprint, contributing to more sustainable industrial practices. On-site hydrogen production supports personalized manufacturing processes, where energy demands can vary significantly based on the type and scale of production activities. This level of customization ensures that each industry can optimize its energy use, leading to increased operational efficiency and cost savings.

The ability to produce and store hydrogen locally also greatly enhances the adaptability of industrial operations. Industries can swiftly respond to fluctuations in energy demand by adjusting their hydrogen production rates accordingly. This responsiveness is particularly valuable in dynamic market environments where demand can change rapidly. For example, during periods of peak demand, industries can ramp up hydrogen production to meet their energy needs without relying on external suppliers. Conversely, during low-demand periods, they can store excess hydrogen for future use, ensuring a steady supply of energy at all times. This adaptability not only improves operational efficiency but also enhances the resilience of industrial processes, reducing the risk of disruptions caused by energy shortages or supply chain issues. Moreover, the local production and storage of hydrogen enable industries to implement more sustainable and resilient energy strategies. By maintaining control over their energy resources, industries can better manage their energy consumption and reduce their dependence on external energy supplies, which are often subject to price volatility and supply disruptions. This self-sufficiency is particularly important in regions with unstable energy markets or limited access to traditional energy sources. Additionally, localized hydrogen production can support the integration of renewable energy sources, such as solar or wind, into industrial energy systems. By using excess renewable energy to produce hydrogen, industries can create a more balanced and sustainable energy mix, further reducing their environmental impact.

Hydrogen production can be decentralized, significantly reducing dependency on fossil fuels and enhancing energy security. This shift toward decentralized energy production is crucial for maintaining stable operations, especially amid the unpredictable fluctuations of global energy markets. By generating hydrogen locally, industries can mitigate the risks associated with the volatility of fossil fuel supplies and prices. This local production reduces the reliance on imported fuels, which can be subject to geopolitical tensions and supply chain disruptions. Consequently, industries can achieve greater energy independence, ensuring a more reliable and predictable energy supply. This energy independence not only stabilizes operational costs but also allows industries to plan and manage their energy needs more effectively.

Furthermore, decentralized hydrogen production contributes to greater resilience and flexibility in energy management. Industries can adjust their hydrogen production according to their specific energy requirements, scaling up or down based on demand. This adaptability ensures that industries are better equipped to handle energy demands without being at the mercy of external fuel supplies. The local generation of hydrogen also supports the integration of renewable energy sources, creating a more sustainable and secure energy ecosystem. By using renewable energy to produce hydrogen, industries can further reduce their carbon footprint and promote a cleaner environment. In addition to enhancing energy independence, hydrogen systems play a vital role in disaster recovery by providing backup power during disruptions. Natural disasters, power outages, and other emergencies can severely impact the availability of electricity, jeopardizing the operation of critical infrastructure and industrial processes. Hydrogen fuel cells and storage systems offer a reliable backup power solution that can be quickly deployed in such situations. These systems ensure that essential services and operations continue to function, thus minimizing the impact of disruptions. For instance, during a power outage, hydrogen fuel cells can provide continuous power to hospitals, data centers, and other critical facilities, ensuring their uninterrupted operation. Hydrogen systems are also crucial for industrial processes that require a consistent power supply. In the event of grid failures or other disruptions, hydrogen backup systems can maintain the operation of key industrial equipment, preventing costly downtimes and production losses. This capability is particularly important for industries that rely on continuous processes, such as chemical manufacturing, where any interruption can lead to significant financial and operational setbacks. Moreover, the use of hydrogen as a backup power source supports the overall resilience of energy systems. By incorporating hydrogen into their energy strategies, industries can diversify their energy sources, reducing the risk of overreliance on a single supply chain. This diversification enhances the robustness of energy systems, making them less vulnerable to disruptions and more capable of recovering quickly from emergencies.

Green hydrogen production and usage are pivotal in achieving net-zero emission targets, positioning hydrogen as a cornerstone for future industries. The drive toward net-zero emissions necessitates the adoption of energy sources that produce no GHGs, and green hydrogen fits this requirement perfectly. Produced through electrolysis powered by renewable energy, green hydrogen is entirely carbon-free, offering a clean alternative to fossil fuels. Industries with processes that are challenging to electrify, such as heavy industrial operations, metal refining, and chemical manufacturing, will increasingly rely on hydrogen. Additionally, long-haul transportation, including trucking, maritime shipping, and aviation, stands to benefit significantly from hydrogen fuel cells and hydrogen combustion engines. These applications highlight hydrogen's versatility and its critical role in decarbonizing sectors where electrification alone is insufficient.

The reliance on hydrogen for net-zero emissions is not just about replacing fossil fuels but it's also about integrating a cleaner energy solution that fits seamlessly into existing industrial frameworks. For example, in steel production, hydrogen can replace coke (a derivative of coal) in the reduction of iron ore, drastically reducing carbon emissions. Similarly, in the chemical industry, hydrogen can serve as a feedstock for producing NH_3 and CH_3OH, essential chemicals for various industrial processes, without the associated carbon emissions of traditional methods. The adoption of hydrogen in these sectors not only helps in meeting emission targets but also ensures the sustainability of industrial operations in the long-term. In addition to its role in achieving net-zero emissions, hydrogen technology significantly supports the circular economy by enabling the reuse of resources and minimizing waste. The concept of a circular economy revolves around designing industrial systems that are restorative and regenerative by intention and design. Hydrogen production and usage align well with these principles. For instance, excess renewable energy, which would otherwise go to waste during periods of low demand, can be utilized to produce hydrogen. This hydrogen can then be stored and used later, effectively balancing supply and demand and reducing energy waste. Moreover, the by-products of hydrogen use, primarily water, pose no environmental harm, unlike the pollutants from fossil fuels. This characteristic of hydrogen makes it an ideal component in a circular economy, where reducing waste and reusing resources are paramount. In industrial processes, the integration of hydrogen can lead to more efficient use of raw materials. For example, hydrogen can be used in closed-loop systems where it is recycled and reused, minimizing the need for fresh inputs and reducing overall resource consumption. The circular economy benefits of hydrogen extend to its potential in creating sustainable industrial ecosystems. Industries can develop symbiotic relationships where the by-products of one process serve as inputs for another. For instance, the oxygen produced as a by-product of electrolysis can be used in wastewater treatment or in industrial processes that require pure oxygen, thereby creating a closed-loop system that enhances resource efficiency.

1.5.4 ROLE OF ARTIFICIAL INTELLIGENCE IN HYDROGEN TECHNOLOGY DEVELOPMENT

Artificial Intelligence (AI) plays a transformative role in advancing hydrogen technology, optimizing production processes, enhancing efficiency, and reducing costs.

The integration of AI into hydrogen technology development accelerates innovation and addresses key challenges in the hydrogen value chain.

1.5.4.1 AI Application in Hydrogen Technology

AI plays a crucial role in enhancing the efficiency and effectiveness of hydrogen production, storage, and distribution processes. One of the primary applications of AI in this field is the optimization of production processes. AI algorithms can analyze extensive datasets related to the electrolysis process used to produce green hydrogen. By doing so, these algorithms can identify patterns and relationships that human analysts might miss, leading to improved energy efficiency and reduced operational costs. Machine learning models can predict the optimal operating conditions for electrolysis, ensuring that hydrogen yield and purity are maximized. This optimization not only makes the production process more efficient but also more cost-effective, making green hydrogen a more viable alternative to traditional fossil fuels.

In addition to optimizing production processes, AI-powered predictive maintenance systems are transforming the way hydrogen production and storage equipment are managed. These systems continuously monitor the equipment, using data from sensors and historical performance records to predict potential failures before they occur. This predictive capability reduces unplanned downtime, which can be costly and disruptive. By addressing maintenance issues proactively, the lifespan of the equipment is extended, and the continuous and reliable supply of hydrogen is ensured. This is particularly important in industrial settings where interruptions in hydrogen supply can halt production and lead to significant financial losses.

AI also significantly improves energy management in hydrogen systems by forecasting energy demand and supply patterns. AI systems can analyze various factors, including historical energy usage data, weather forecasts, and market trends, to predict future energy needs. This predictive capability ensures that the production, storage, and distribution of hydrogen are aligned with actual demand, optimizing the use of resources and enhancing grid stability. By efficiently integrating hydrogen with renewable energy sources, AI systems help to balance supply and demand, preventing overproduction and reducing waste. This optimization is crucial for maintaining a stable and reliable energy grid, especially as the share of intermittent renewable energy sources like wind and solar continues to grow.

Furthermore, AI's role in energy management extends to real-time adjustments of energy flows, ensuring that hydrogen is produced when renewable energy is plentiful and stored for use when demand is high or renewable generation is low. This dynamic management of energy resources enhances the overall efficiency of the energy system, making it more resilient and capable of adapting to fluctuations in both supply and demand.

1.5.4.2 Enhancing Safety and Efficiency

AI significantly enhances safety in hydrogen production, storage, and transportation by providing continuous monitoring for potential hazards such as leaks, pressure changes, and other anomalies. Using advanced sensors and data analysis, AI systems can detect deviations from normal operating conditions in real time. This continuous surveillance allows for immediate responses to potential safety issues, preventing accidents before they occur. For example, if a leak is detected, AI-driven safety protocols can automatically shut down the affected area and alert maintenance teams, minimizing the risk of explosions or other hazardous events. By maintaining a constant vigil over the hydrogen infrastructure, AI ensures a safer environment for workers and the surrounding community.

In addition to safety monitoring, AI plays a crucial role in optimizing the hydrogen supply chain. By predicting demand and managing logistics, AI ensures that hydrogen is produced, stored, and delivered efficiently. Advanced algorithms analyze market trends, historical data, and real-time information to forecast hydrogen demand accurately. This predictive capability allows producers to adjust their output to match demand, reducing overproduction and waste. Furthermore, AI optimizes transportation routes and schedules, ensuring that hydrogen reaches its destination in a timely manner. This logistical efficiency not only reduces costs but also enhances the reliability of hydrogen as an energy carrier, making it a more attractive option for various applications.

AI also accelerates R&D in hydrogen technology by analyzing vast amounts of scientific data to identify new materials and processes. Traditionally, R&D in this field has been time-consuming and resource-intensive, but AI can significantly speed up this process. By sifting through data from experiments, simulations, and academic publications, AI can identify patterns and correlations that might be overlooked by human researchers. This capability leads to the discovery of more efficient catalysts, which are crucial for improving the efficiency of hydrogen production processes like electrolysis. Additionally, AI helps in developing better storage materials that can hold hydrogen more safely and compactly. In fuel cells, AI-driven research can lead to designs that are more efficient, durable, and cost-effective, thus advancing the overall technology.

REFERENCES

1. Bareiß, K., et al., *Life cycle assessment of hydrogen from proton exchange membrane water electrolysis in future energy systems.* Applied Energy, 2019. **237**: p. 862–872.
2. Nikolaidis, P. and A. Poullikkas, *A comparative overview of hydrogen production processes.* Renewable and sustainable energy reviews, 2017. **67**: p. 597–611.
3. Arsad, A., et al., *Hydrogen electrolyser technologies and their modelling for sustainable energy production: A comprehensive review and suggestions.* International Journal of Hydrogen Energy, 2023. **48**(72): p. 27841–27871.
4. Veziroglu, T., *Hydrogen Energy: Part A.* 2012: Springer Science & Business Media.
5. Veziroglu, T.N., S. Sherif, and F. Barbir, *Hydrogen energy solutions*, in *Environmental solutions.* 2005, Elsevier. p. 143–180.
6. Pal, A., et al., *Powering squarely into the future: A strategic analysis of hydrogen energy in QUAD nations.* International Journal of Hydrogen Energy, 2023. **49**: p. 16–41.
7. Zainal, B.S., et al., *Recent advancement and assessment of green hydrogen production technologies.* Renewable and Sustainable Energy Reviews, 2024. **189**: p. 113941.
8. Guilbert, D. and G. Vitale, *Hydrogen as a clean and sustainable energy vector for global transition from fossil-based to zero-carbon.* Clean Technologies, 2021. **3**(4): p. 881–909.
9. Koroneos, C., et al., *Life cycle assessment of hydrogen fuel production processes.* International journal of hydrogen energy, 2004. **29**(14): p. 1443–1450.

10. Grahame, A. and K.-F. Aguey-Zinsou, *Properties and Applications of Metal (M) dodecahydro-closo-dodecaborates (Mn= 1, 2B12H12) and Their Implications for Reversible Hydrogen Storage in the Borohydrides*. Inorganics, 2018. **6**(4): p. 106.

11. Zhang, M. and X. Yang, *The Regulatory Perspectives to China's Emerging Hydrogen Economy: Characteristics, Challenges, and Solutions*. Sustainability, 2022. **14**(15): p. 9700.

12. Iida, S. and K. Sakata, *Hydrogen technologies and developments in Japan*. Clean Energy, 2019. **3**(2): p. 105–113.

13. Włodarczyk, R. and P. Kaleja, *Modern hydrogen technologies in the face of climate change—Analysis of strategy and development in Polish conditions*. Sustainability, 2023. **15**(17): p. 12891.

14. Sedai, A., et al., *Wind energy as a source of green hydrogen production in the USA*. Clean Energy, 2023. **7**(1): p. 8–22.

15. Qazi, U.Y., *Future of hydrogen as an alternative fuel for next-generation industrial applications; challenges and expected opportunities*. Energies, 2022. **15**(13): p. 4741.

2 Hydrogen Production

2.1 INTRODUCTION TO HYDROGEN PRODUCTION

Hydrogen production is essential for the global energy transition, offering a sustainable, low-carbon future. It can decarbonize sectors resistant to electrification, such as heavy industry, long-haul transportation, and building heating. Additionally, hydrogen aids in storing and transmitting intermittently generated renewable energy, stabilizing systems reliant on solar and wind power. SMR is the predominant industrial method, reacting CH_4 with steam to produce hydrogen and CO_2. It has high energy efficiency, typically between 70% and 85%, but generates significant CO_2 emissions unless coupled with CCS technologies. SMR is currently the most economically viable method. Water electrolysis, especially when powered by renewable electricity, has a minimal carbon footprint and an energy efficiency ranging from 60% to 80%. However, it is generally more expensive due to high electricity costs. Coal gasification converts coal into a hydrogen-rich gas mixture but has lower energy efficiency, often between 40% and 55%, and similar CO_2 emission challenges.

Currently, most hydrogen is produced from fossil fuels, leading to significant CO_2 emissions. However, interest in "green" hydrogen, produced via water electrolysis using renewable electricity, is growing. This method emits almost no CO_2, with water being its only by-product. The European Union aims to install 6 GW of renewable hydrogen electrolyzers by 2024 and 40 GW by 2030, reflecting a global trend toward increasing the share of hydrogen produced from renewable sources. Hydrogen also plays a crucial role in decarbonizing energy-intensive industries like steelmaking, chemicals, and cement production by replacing traditional fossil fuels. In transportation, hydrogen FCEVs offer a solution for reducing emissions from heavy-duty vehicles. Additionally, hydrogen facilitates long-term renewable energy storage and can be integrated into existing energy networks to enhance flexibility and efficiency.

Advancements in hydrogen production technologies are essential for overcoming challenges related to cost, efficiency, and scalability. Innovations in electrolysis, CCS, and new catalysts and materials are critical for reducing production costs and improving efficiency. International collaborations and significant investments in research and development drive these technological advancements. Consequently, hydrogen production is at the forefront of the global energy transition, providing a versatile and sustainable solution for reducing carbon emissions across various sectors. Continuous improvements in production technologies, supportive policies, and investments in renewable energy infrastructure will be crucial for realizing hydrogen's full potential in a low-carbon future.

DOI: 10.1201/9781032718453-2

2.1.1 Fundamentals and Context of Hydrogen Production

2.1.1.1 Energy Efficiency of Hydrogen Production

Energy efficiency in hydrogen production varies significantly across different methods. SMR, the predominant industrial method, has an energy efficiency ranging from 70% to 85%, though it generates significant CO_2 emissions unless CCS technologies are employed. Water electrolysis splits water molecules into hydrogen and oxygen using electrical energy, typically achieving an efficiency rate between 60% and 80%, influenced by the electricity source and electrolyzer efficiency. Coal gasification, which converts coal into a hydrogen-rich gas mixture, has lower energy efficiency, often between 40% and 55%, and similar CO_2 emission challenges. While electrolytic hydrogen production using renewable electricity has a minimal carbon footprint, its energy efficiency is generally lower than SMR or coal gasification. Nevertheless, hydrogen's potential as a clean energy carrier, especially from renewable sources, and its versatility across applications in transportation, energy storage, and industrial processes are significant. A comprehensive evaluation of hydrogen production methods must consider energy efficiency, environmental impacts, economic viability, resource availability, and safety.

2.1.1.2 The Share of Hydrogen Produced from Renewable Sources

The share of hydrogen produced from renewable sources, such as solar and wind energy, is currently low but is expected to increase significantly in the coming years. At present, the majority of hydrogen is produced from fossil fuels, particularly through SMR, which leads to significant CO_2 emissions. However, there is a growing interest in "green" hydrogen, produced via water electrolysis using electricity sourced from renewable sources like wind and solar energy.

Currently, there are two primary methods for producing green hydrogen using seawater: one-step and two-step electrolysis, as illustrated in Figure 2.1. In the two-step electrolysis of seawater, water is first filtered by a reverse osmosis membrane module and then subjected to electrolysis. Conversely, in the one-step electrolysis of seawater, water is electrolyzed immediately after a simple pretreatment [1]. This method of hydrogen production is considered the cleanest, as it emits almost no CO_2, with water being its only by-product.

The EU, in particular, is emphasizing the development and expansion of renewable hydrogen production to leverage its flexibility and versatility in driving the transition toward clean energy. The EU's hydrogen strategy aims to install at least 6 GW of renewable hydrogen electrolyzers by 2024 and 40 GW by 2030. This aligns with broader efforts to decarbonize energy-intensive industries and transportation sectors that are not easily electrifiable.

It's also noteworthy that the International Energy Agency (IEA) and the International Renewable Energy Agency (IRENA) estimate that green hydrogen could meet 12–13% of final energy demand by 2050, compared to nearly zero today. This indicates significant growth potential for green hydrogen in the coming decades, although numerous challenges remain to be addressed, including technological improvements, cost competitiveness, and international policy implications. In summary, while the share of hydrogen produced from renewable sources is currently

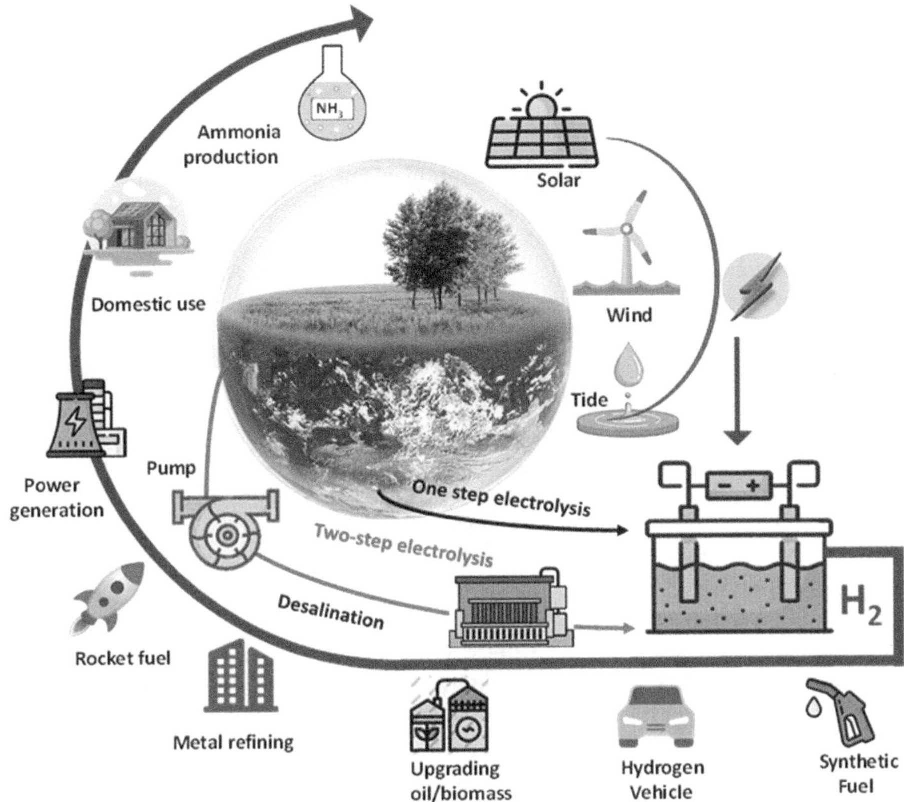

FIGURE 2.1 Two distinct strategies for extracting hydrogen from seawater using alternative power sources.

(Adapted from ref. [2].)

limited, ongoing plans and investments signal a promising future for green hydrogen as a key component of the global energy transition toward cleaner and more sustainable systems.

2.1.1.3 The Contribution of Hydrogen Production to the Transition to a Low-Carbon Economy

Hydrogen, as a versatile energy carrier, holds promise in catalyzing the transition toward a low-carbon economy through several pivotal avenues. First and foremost, its role in decarbonizing energy-intensive industries like steelmaking, chemicals, and cement production cannot be overstated. By substituting traditional fossil fuels with hydrogen, particularly green hydrogen derived from renewable sources, these sectors can significantly mitigate their carbon footprint, contributing to overall emission reductions on a substantial scale.

Additionally, hydrogen presents a viable solution for the long-term storage of renewable energy, addressing the intermittency challenge inherent in sources like solar and wind power. Through electrolysis, excess renewable energy can be harnessed to produce hydrogen, which can then be stored and utilized during periods of high demand or low renewable energy availability. This not only bolsters grid stability but also enhances the integration of renewable energy into the broader energy mix, fostering a more sustainable and resilient energy system.

Furthermore, hydrogen's role in enabling clean mobility, particularly through FCEVs, offers a promising avenue for decarbonizing the transportation sector. Heavy-duty vehicles such as trucks and buses, which pose significant challenges for direct electrification, can benefit greatly from hydrogen-powered propulsion systems, eliminating direct emissions of pollutants and CO_2 and contributing to cleaner air quality and reduced GHG emissions.

Lastly, hydrogen serves as a key enabler of energy system flexibility by acting as an intermediary between different energy sectors, including gas, electricity, and heat. This facilitates the optimization of energy flows, enhances overall system efficiency, and fosters the transition toward a more interconnected and adaptive low-carbon energy ecosystem. However, realizing hydrogen's full potential necessitates overcoming existing technological, economic, and infrastructural barriers, while ensuring that its production processes adhere to sustainable and environmentally responsible practices. A summary of hydrogen production contribution to the transition to a low-carbon economy is outlined in Table 2.1.

The primary energy sources for the world over the past two centuries have been crude oil and coal. Modern society is heavily reliant on these energy sources. However, the finite availability of these resources, combined with increasing energy demand, has created significant pressure on their usage. Additionally, the combustion of fossil fuels leads to a rise in CO_2 emissions, which amounted to 266,652 metric tons in 2016, particularly from the energy and transport sectors. Reducing carbon emissions is essential

---

TABLE 2.1

Summary of Hydrogen Production Contribution

Key Contribution	Description
Decarbonization of industries	Utilization of hydrogen in sectors like steel, chemicals, and cement to reduce CO_2 emissions from fossil fuel reliance
Renewable energy storage	Conversion of excess solar and wind power into hydrogen for long-term storage, aiding grid stability and renewable integration
Clean transportation	Adoption of hydrogen FCEVs in challenging sectors for emission reduction
Energy system flexibility	Utilization of hydrogen as an energy carrier to enhance system efficiency and flexibility during the low-carbon transition
Reduced fossil fuel reliance	Transition of various energy applications from fossil fuels to hydrogen to diminish dependency and emissions
Green economy development	Expansion of green hydrogen sectors stimulating economic growth, innovation, and job opportunities
Decarbonization in heating and cooking	Integration of hydrogen into natural gas networks to decrease CO_2 emissions from heating and cooking, facilitating a smoother transition to clean energy sources

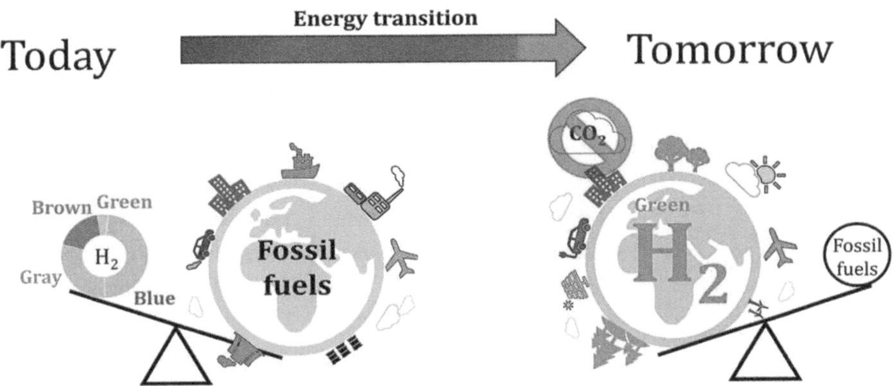

FIGURE 2.2   Global energy system transition.

(Adapted from ref. [3].)

to mitigate the risk of global warming caused by GHGs. Concerns about energy supply security and environmental issues have also prompted a shift toward alternative energy forms, including solids, liquids, and gases, as illustrated in Figure 2.2. By 2080, hydrogen is predicted to constitute 90% of the global energy supply [3].

2.1.1.4 Countries Leading in Hydrogen Production

Hydrogen, as an energy carrier, is produced by several countries around the world, each using different production methods and having varying levels of output. The United States is one of the largest hydrogen producers globally. Production is primarily achieved through SMR, but efforts are underway to increase green hydrogen production from renewable sources. The government and private enterprises are investing heavily in research and development to make this production more sustainable and less dependent on fossil fuels.

The leading countries in hydrogen production encompass a variety of global players that have adopted national strategies to develop this energy resource. Among them, China stands out as the primary producer and consumer of hydrogen since 2010, with a notable production mainly from coal gasification. However, China is considering transitioning to hydrogen production via electrolysis powered by renewable sources, which could represent 80% of its supply by 2060. This shift toward green hydrogen aligns with China's broader goal of achieving carbon neutrality by 2060, despite the country's current dependence on coal for hydrogen production

In Europe, several countries have developed ambitious hydrogen strategies, including France, Germany, and others like Belgium, Denmark, Spain, and Italy, each with its own objectives and action plans to integrate hydrogen into their energy mix. For instance, France is focusing on building a comprehensive hydrogen ecosystem, while Germany is concentrating on the use of renewable hydrogen produced from photovoltaic electricity in sunny regions like the Maghreb or Gulf countries.

Beyond Europe and Asia, other countries, including Chile and Morocco, as well as developed nations like Canada and the United States, have also established strategies

TABLE 2.2

Quantity of Hydrogen Produced by Each Country and Their Main Usage Domains

Country	Hydrogen Production (tons/year)	Main Usage Domains
China	33,000,000	Industry, Energy Production, Transportation
United States	11,000,000	Industry, Transportation, Energy Storage
Russia	5,000,000	Industry, Energy Production, Transportation
Canada	3,000,000	Industry, Transportation, Energy Export
Germany	1,700,000	Industry, Transportation, Energy Storage
South Korea	1,500,000	Transportation, Industry, Energy Storage
Japan	1,000,000	Transportation, Energy Storage, Industry
France	900,000	Industry, Transportation, Energy Storage
Netherlands	250,000	Industry, Transportation, Energy Export
Australia	200,000	Energy Export, Industry, Transportation
Norway	100,000	Energy Storage, Industry, Transportation

for hydrogen development, particularly green hydrogen. These national efforts range from massive renewable hydrogen production for export to creating domestic markets for hydrogen as a clean energy vector.

China, in particular, made significant progress in 2023, substantially increasing its production and use of green hydrogen. The country deployed a provincial-scale strategy to develop hydrogen production, with plans established in 27 provinces to accelerate production. China also exceeded its own forecasts for hydrogen vehicles, with a revised target of 117,000 hydrogen vehicles on the road by 2025. These developments reflect a growing global trend toward adopting hydrogen as a key element in the energy transition, with leading countries implementing ambitious strategies to harness this clean and sustainable energy source.

Table 2.2 provides an overview of the production volumes and primary uses of hydrogen in these countries. It provides an illustrative overview of hydrogen production capacities and their uses in different countries. The figures and usage domains are approximations based on general trends observed in the hydrogen sector.

2.1.2 ADVANCES IN TECHNOLOGY AND INNOVATION

2.1.2.1 Recent Progress in Hydrogen Production Technologies

Recent advancements in hydrogen production technologies are primarily focused on improving efficiency and reducing costs, with a particular emphasis on decarbonized and green hydrogen production. In France, the national strategy for the development of decarbonized hydrogen, announced in 2020, aims to accelerate the deployment of hydrogen as a key element in industrial decarbonization. France plans to invest €9 billion in developing the renewable and low-carbon hydrogen sector, with the goal of installing 6.5 GW of electrolysis capacity by 2030. This would represent the production of 600 kt/year of decarbonized hydrogen, thus promoting ecological transition and creating a dedicated industrial sector. This initiative includes funding for research and development, deployment of hydrogen-powered trains, and support for innovation in electrolysis technologies to improve energy efficiency and reduce costs.

Internationally, the Global Hydrogen Production Technologies (HyPT) Center establishes an international partnership involving six countries (the United States, Australia, Canada, the United Kingdom, Egypt, and Germany) to formulate a pathway toward large-scale, low-cost net-zero hydrogen production. The goal is to achieve net-zero hydrogen production for $1/kg at a gigaton scale per year. This center is led by several leading academic institutions and receives support from various national funding agencies, reflecting a collaborative effort to overcome challenges related to hydrogen production from renewable and low-carbon sources.

These efforts represent a significant step toward reducing dependence on fossil fuels for hydrogen production, focusing on efficiency, cost reduction, and integration of renewable energies into the production process. International collaboration and investments in research and development are crucial to achieving these goals and making hydrogen a key component of the global energy transition.

2.1.2.2 Future Innovations for Sustainable and Efficient Hydrogen Production

Future innovations in the field of hydrogen production aim to make it more sustainable, efficient, and economically viable. Several promising research and innovation directions include enhancing the efficiency of electrolyzers, integrating renewable energies directly into hydrogen production processes, exploring bioproduction methods using microorganisms, capturing hydrogen as a by-product from industrial processes, developing advanced hydrogen storage technologies, exploring high-temperature electrolysis (HTE), and researching water photolysis.

Improving the efficiency of electrolyzers is crucial for reducing the cost of hydrogen production by water electrolysis. This involves developing new catalysts, utilizing more efficient membranes, and integrating CO_2 capture technologies for processes using fossil fuels. Another key goal is producing green hydrogen from renewable energies like solar, wind, and hydroelectric power. Directly integrating electrolyzers with renewable energy sources can lead to more efficient and cost-effective hydrogen production. Research into bioproduction methods, such as using algae or bacteria to produce hydrogen under certain conditions, shows promise for generating hydrogen from renewable resources without requiring external electrical energy. Capturing and utilizing hydrogen produced as a by-product in various industrial processes, like chemical industries or oil refining, presents an opportunity to increase hydrogen supply without significantly increasing CO_2 emissions.

Moreover, advancements in hydrogen storage technologies, such as advanced metal hydrides or nanoporous materials, can revolutionize the hydrogen economy by making storage more efficient and less costly. HTE, although still experimental, holds promise for significantly reducing hydrogen production costs by using less electricity. Research into water photolysis, which uses sunlight to directly split water into oxygen and hydrogen without an electrolyzer, could offer a fully renewable and highly efficient hydrogen production method. Continuous innovation in these areas and others is crucial for overcoming current challenges in hydrogen production and realizing its potential as a key energy carrier for a low-carbon economy.

2.1.3 Utilization and Integration of Hydrogen

2.1.3.1 The Main Applications of Hydrogen Once Produced

Hydrogen, once produced, exhibits a diverse range of applications that underscore its pivotal role in the global energy transition toward a low-carbon economy. Scientifically speaking, these applications are underpinned by the unique properties of hydrogen as an energy carrier and its potential to address various challenges associated with traditional energy sources.

One significant application lies in mobility and transport, where hydrogen serves as a clean and efficient energy source for FCEVs. They convert hydrogen gas into electricity through a chemical reaction with oxygen, emitting only water vapor as a by-product. Advancements in fuel cell technology, hydrogen storage, and infrastructure development are essential to enhancing the performance and widespread adoption of FCEVs. Additionally, hydrogen plays a vital role in energy storage, particularly in balancing the intermittency of renewable energy sources like wind and solar. Through electrolysis, surplus electricity from renewables can be used to split water molecules into hydrogen and oxygen, which can then be stored for later use. Advanced research in electrolysis technology, hydrogen compression, and storage materials is crucial in optimizing the efficiency and scalability of hydrogen storage systems.

Moreover, hydrogen serves as a versatile feedstock in industrial processes, including petroleum refining, ammonia production, and metallurgy. Green hydrogen produced from renewable sources offers a sustainable alternative to hydrogen derived from fossil fuels, thereby reducing carbon emissions across various industrial sectors. Scientific innovations in electrolysis efficiency, catalyst development, and process optimization are critical for advancing the green hydrogen economy.

Lastly, hydrogen holds promise as a clean energy source for residential and commercial heating applications, contributing to efforts to reduce GHG emissions in buildings. Hydrogen boilers and blended hydrogen–natural gas systems can help decarbonize heating systems, with ongoing scientific research into combustion characteristics, heating system compatibility, and safety standards essential to effective integration. These scientific developments underscore hydrogen's multifaceted role as a versatile and sustainable energy carrier, driving innovation across various sectors and contributing to the global transition toward a more resilient and carbon-neutral energy system.

2.1.3.2 Integration of Hydrogen Production into Existing Energy Networks

The integration of hydrogen production into existing energy networks represents a complex endeavor with profound scientific and technological ramifications. One fundamental advantage lies in its role as a storage medium for renewable energy. Through electrolysis, surplus electricity generated from renewable sources during periods of low demand, such as from wind or solar power, can be harnessed to produce hydrogen. This hydrogen reservoir can then be stored and utilized later, serving as a mechanism to mitigate disparities between energy supply and demand within the electrical grid. Moreover, hydrogen's versatility extends to its function as a flexible energy carrier, facilitating its reconversion into electricity during periods of peak demand or reduced renewable energy availability through fuel cells or gas turbines.

Furthermore, hydrogen holds promise as an efficient means of energy transport and distribution. By leveraging existing natural gas infrastructure with suitable adjustments, hydrogen can be efficiently transported and disseminated. This adaptation obviates the need for constructing entirely new energy transmission infrastructure, thereby minimizing integration costs. Additionally, hydrogen can be blended with natural gas and conveyed through existing networks, contributing to the mitigation of carbon emissions in various sectors, including heating, electricity generation, and industrial processes. In industrial and transportation decarbonization, hydrogen emerges as a pivotal solution. It can supplant fossil fuels in high-temperature industrial applications such as steelmaking and refining, where direct electrification is often unfeasible or economically unviable. Similarly, within the transportation sector, hydrogen serves as a viable fuel option for fuel cell vehicles, particularly in heavy-duty transport and maritime industries, where challenges related to weight and range constraints impede direct electrification endeavors. Furthermore, the integration of hydrogen catalyzes innovation and propels the energy transition forward. It fosters the development of new energy markets and services, facilitating the emergence of clean mobility solutions, grid-balancing mechanisms, and decarbonized heating systems. Investment in hydrogen-related research, development, and deployment not only expedites the adoption of hydrogen technologies but also drives advancements in critical scientific domains essential for realizing broader energy transition objectives.

2.1.4 CHALLENGES, POLICIES, AND ENVIRONMENTAL IMPACTS

2.1.4.1 Environmental Challenges Associated with Hydrogen Production

Hydrogen production poses significant environmental challenges, with its environmental impact varying depending on the method used. Conventional methods, such as steam reforming of natural gas, are associated with substantial CO_2 emissions. Even with CCS technologies to produce "blue" hydrogen, concerns persist regarding the efficiency and safety of CO_2 storage. Green hydrogen production via electrolysis, while emissions-free at the point of use, demands substantial electricity, and if sourced from nonrenewable sources, can indirectly contribute to GHG emissions. Therefore, the choice of production method is critical in determining the environmental footprint of hydrogen. Water usage is another critical concern in hydrogen production, particularly with electrolysis, which consumes significant amounts of freshwater. In regions already facing water scarcity, this could exacerbate stress on water resources. Additionally, improper management of electrolysis processes can result in water quality degradation. The development of large-scale renewable energy infrastructure for green hydrogen production also raises concerns about its impact on local ecosystems and biodiversity, including habitat disruption and alteration of natural landscapes. Assessing and mitigating these impacts are vital aspects of sustainable hydrogen production.

Furthermore, hydrogen production processes may generate waste that requires proper handling and disposal. Chemical processes and the use of specific materials can result in the production of industrial waste, necessitating environmentally sound waste management practices. Additionally, the construction and operation of infrastructure for hydrogen production, storage, and distribution can have environmental ramifications, including resource consumption and the generation of additional industrial waste. To address these challenges, comprehensive approaches that prioritize energy efficiency, promote the use of renewable energy sources, minimize water usage, and implement sustainable waste management practices are essential. Continued research, innovation, and policy development are crucial for advancing environmentally sustainable hydrogen production methods and mitigating its environmental impacts effectively.

2.1.4.2 Technical and Economic Challenges

The production of hydrogen, particularly through green pathways like electrolysis powered by renewable energy sources, confronts a spectrum of intricate technical challenges demanding innovative solutions. Enhancing the efficiency of electrolysis processes stands as a paramount objective, necessitating advancements in catalyst materials, membrane technologies, and system designs to minimize energy losses and optimize hydrogen yields. Material durability emerges as a critical concern, particularly in the context of electrolyzer components exposed to harsh operating conditions, necessitating the development of corrosion-resistant materials and robust maintenance protocols to ensure prolonged equipment lifespans. Moreover, the intricate logistics of hydrogen storage and transport present formidable hurdles due to its low energy density, driving research into novel storage methods like hydrogen carriers and innovative transport solutions to facilitate large-scale deployment and distribution.

On the economic front, the transition to hydrogen faces formidable cost barriers that must be addressed to ensure its competitiveness in the energy landscape. Chief among these is the high cost of electrolysis technologies, which demands concerted research efforts to drive down capital expenditures and operational costs through technological advancements and economies of scale. Significant investments in infrastructure are imperative to establish a comprehensive hydrogen ecosystem encompassing production facilities, storage infrastructure, distribution networks, and refueling stations, requiring strategic public–private partnerships and targeted government incentives to mobilize capital and mitigate investment risks. Additionally, ensuring equitable access to abundant and affordable renewable energy sources is pivotal for scaling up green hydrogen production, underscoring the importance of fostering robust renewable energy markets and incentivizing renewable energy deployment to support the hydrogen transition.

Technical challenges in hydrogen production include improving the efficiency of electrolysis, enhancing the durability of materials used, ensuring the safe and effective storage and transport of hydrogen, and integrating hydrogen into existing energy networks. Electrolysis, a key process in hydrogen production, requires advancements to become more energy efficient. Material durability is crucial to withstand the harsh conditions of hydrogen production and usage. Additionally, hydrogen storage and transport present significant technical hurdles due to its low density and high flammability. Successfully integrating hydrogen into energy networks also demands sophisticated technology to manage the variability and distribution of energy supply.

Economic challenges are equally significant, with the high cost of electrolysis being a major barrier. Infrastructure investments are necessary to build and maintain the facilities required for hydrogen production, storage, and distribution. Access to renewable energy sources is critical for producing green hydrogen, which can be cost-prohibitive in certain regions. Furthermore, hydrogen faces competition from other energy vectors, making it essential to establish a competitive market position. Balancing these economic factors is essential to making hydrogen a viable and sustainable energy solution.

2.1.4.3 Government Policies

Government policies play a pivotal role in shaping the landscape of hydrogen production, particularly in the advancement of clean hydrogen technologies. These policies are designed to incentivize investment, stimulate innovation, and create an enabling environment for the growth of the hydrogen industry. One key aspect of government intervention lies in providing financial support through subsidies and grants for research and development initiatives related to hydrogen production technologies. Direct funding encourages private sector participation in the development of innovative solutions, such as electrolysis methods and low-carbon hydrogen production techniques, which are essential for the transition toward sustainable hydrogen production. Tax incentives and fiscal benefits are another effective policy tool employed by governments to promote hydrogen production. Tax credits for companies investing in green hydrogen infrastructure or purchasing hydrogen-related equipment serve as financial incentives to spur investment in the sector. Additionally, tax deductions and relief on operational and capital costs associated with clean hydrogen production and utilization help reduce the financial burden on industry players, making hydrogen production economically more viable.

Regulatory standards and objectives play a crucial role in creating a conducive environment for the adoption of hydrogen technologies. Governments often set decarbonization targets and introduce blending standards that mandate the inclusion of hydrogen in the natural gas mix, thereby increasing demand for hydrogen and encouraging market growth. Furthermore, initiatives such as emission reduction goals in energy, transportation, and industrial sectors provide a clear direction for industry players and drive innovation toward cleaner hydrogen production methods.

International cooperation and commitment are also essential components of government policies aimed at promoting hydrogen production. Participation in international initiatives and agreements fosters knowledge sharing, technology transfer, and collaboration in research and development efforts. By engaging in platforms such as the Hydrogen Energy Ministerial Meeting and the Hydrogen Initiative under Mission Innovation, governments demonstrate their commitment to advancing the global transition toward clean hydrogen, ensuring alignment with broader sustainability objectives on a global scale.

Figure 2.3 presents an example of the hydrogen policies and strategies implemented by various countries in the year 2020–2021. These countries are at different stages of implementing hydrogen policies, including R&D programs, vision documents, roadmaps, and strategies. Notably, during this period, several countries developed mature strategy proposals. These proposals vary in their focus (such as green

FIGURE 2.3 Radar map of countries classified by recent hydrogen policies and strategies.
(Adapted from ref. [4].)

hydrogen, fossil-based hydrogen, or a mix of both) and scale, ranging from having no targets to setting very ambitious, specific hydrogen and electrolyzer targets.

2.1.4.4 Optimizing Hydrogen Production to Reduce Carbon Footprint

To optimize hydrogen production and reduce its carbon footprint, a multifaceted approach is required, encompassing technological advancements, policy interventions, and strategic collaborations across various sectors. At the forefront of these efforts is the imperative to shift toward sustainable production methods, particularly emphasizing the scaling up of green hydrogen production. Green hydrogen, produced through electrolysis powered by renewable energy sources, stands out as a key enabler for decarbonizing the hydrogen supply chain. By leveraging solar, wind, or hydroelectric energy, green hydrogen production eliminates CO_2 emissions associated with conventional methods, offering a pathway toward a cleaner and more sustainable energy future.

In parallel, the transition to blue hydrogen, coupled with CCS technologies, presents another avenue for reducing the carbon intensity of hydrogen production. By capturing and storing CO_2 emissions generated during hydrogen production from fossil fuels, such as steam reforming of natural gas, blue hydrogen mitigates its environmental impact. Optimization of CCS technologies and process efficiencies is essential to maximize the carbon abatement potential of blue hydrogen, thereby complementing efforts toward decarbonization.

Investment in research and innovation plays a pivotal role in advancing hydrogen production technologies and driving down costs. Exploring novel approaches such as water photolysis, bioproduction of hydrogen, and advanced catalysis can lead to breakthroughs in efficiency and affordability. Additionally, hybrid systems that integrate multiple renewable energy sources offer promise in enhancing the reliability and resilience of hydrogen production. These advancements underscore the importance of ongoing collaboration between academia, industry, and government institutions to accelerate the development and deployment of sustainable hydrogen solutions.

Furthermore, optimizing resource use and integrating hydrogen production into existing energy networks are critical steps in enhancing the overall sustainability of hydrogen production. Sustainable water management practices, alongside the adoption of durable and low-carbon materials, help minimize environmental impacts. Integration of hydrogen into energy networks facilitates the efficient storage and distribution of renewable energy, bolstering grid stability and enabling greater penetration of intermittent renewables. By implementing these strategies in a coordinated manner, stakeholders can collectively drive toward a more sustainable and decarbonized hydrogen economy, paving the way for a cleaner energy future.

2.1.4.5 Potential Risks

Hydrogen, hailed as a promising energy carrier, presents a spectrum of potential safety hazards that demand meticulous management throughout its lifecycle, from production to utilization. Chief among these risks is its highly flammable nature. Hydrogen can readily ignite in air within a wide range of concentrations, making leaks a significant concern due to the potential for fires or explosions. Moreover, its often high-pressure storage, particularly in transportation applications, introduces the risk of explosions if pressure regulation fails or if containers are compromised.

Another critical safety consideration is temperature. Hydrogen is commonly stored in liquid form at extremely low temperatures, necessitating specialized equipment to mitigate the risks associated with handling substances at such extreme cold temperatures, such as cold burns. Additionally, while hydrogen itself is not toxic, significant leaks can lead to a reduction in oxygen concentration in confined spaces, posing a risk of asphyxiation.

The detectability of hydrogen poses a distinct challenge to safety protocols. Being both colorless and odorless, its detection without specialized equipment is challenging. Consequently, the deployment of hydrogen leak detection systems becomes imperative in facilities where hydrogen is produced or stored to promptly identify and address any leaks. Moreover, the dispersion of hydrogen, facilitated by its small molecular size, compounds this challenge, as it can quickly permeate containment systems, both an advantage and a potential hazard depending on the circumstances.

Furthermore, the risk of thermal runaway, particularly in applications like fuel cells, underscores the importance of effective management of hydrogen-related processes. If not carefully controlled, chemical reactions involving hydrogen can lead to overheating and subsequent equipment failure. Additionally, hydrogen's potential to cause hydrogen embrittlement in certain metals raises concerns regarding the integrity of storage vessels and pipelines, emphasizing the need for rigorous material compatibility assessments and preventive measures. In conclusion, comprehensive safety protocols, including robust leak detection systems, stringent adherence to standards and regulations, specialized equipment, and thorough personnel training, are essential to mitigate the inherent risks associated with hydrogen throughout its lifecycle.

2.1.5 Market, Industrial Players, and Future Perspectives

2.1.5.1 Major Industrial Players

Hydrogen production involves several major players worldwide, operating in different segments of the industry, from initial hydrogen production to its storage, transportation, and final use.

Air Liquide, a French industrial gases leader, produces hydrogen through methods like water electrolysis and natural gas reforming, while also developing renewable hydrogen infrastructure. Linde, a German multinational, focuses on hydrogen production, storage, and distribution, particularly in green hydrogen projects. In the United States, Air Products and Chemicals innovates in hydrogen technology, including hydrogen refueling stations and renewable hydrogen production. Siemens Energy from Germany specializes in green hydrogen production using renewable energy for carbon-free hydrogen via water electrolysis. French energy company ENGIE invests in renewable hydrogen from green energy sources through similar projects. Anglo-Dutch giant Shell is developing hydrogen as an alternative fuel, establishing refueling stations and producing green hydrogen. ITM Power in the UK and NEL Hydrogen from Norway are leaders in electrolysis technologies, enhancing renewable hydrogen production for energy and transportation. These companies are integral to advancing hydrogen as an energy carrier, aiming to cut production costs, enhance efficiency, and minimize environmental impacts, crucial for transitioning to a low-carbon economy.

2.1.5.2 Main Markets

The main markets for hydrogen are undergoing significant evolution, with a focus on green hydrogen to achieve decarbonization goals and meet future global energy needs. Hydrogen is considered essential for decarbonizing sectors that are difficult to electrify directly, such as aviation, maritime transport, and high-temperature manufacturing activities. However, its development in the global energy landscape is deemed slow compared to the targets of the Paris Agreement. It is estimated that hydrogen could account for 0.5% of global final energy consumption by 2030 and 5% by 2050. Europe, thanks to favorable policies, could see hydrogen represent 11% of its final energy consumption by 2050. The main consuming industries of hydrogen will be those using high-temperature manufacturing processes, while its use in electricity production will be limited. Nearly 72% of hydrogen production in 2050 is projected to be green, requiring a surplus of renewable energy and significant electrolyzer capacity.

According to IRENA, to achieve a scenario where global warming remains below 1.5°C, hydrogen and its derivatives should cover 14% of global final energy consumption by 2050. Currently, the majority of global hydrogen production comes from fossil fuels, contributing to climate change. It is urgent to transition this sector toward a clean hydrogen supply, with global hydrogen production needing to increase by over five times by 2050. This will involve a significant expansion of renewable energy capacities and electrolyzer capacities. International trade in green hydrogen and its derivatives will play a crucial role in linking low-cost production areas to high-demand areas, requiring the development of new supply chains and harmonization of regulatory frameworks to encourage the green hydrogen market.

These outlooks underscore the importance of hydrogen, particularly green hydrogen, as a crucial future energy vector for the transition to a low-carbon economy. The development of favorable policies, reliable infrastructure, and international harmonization of standards and regulations are essential to realize the potential of hydrogen in reducing CO_2 emissions and meeting future energy needs sustainably.

2.1.5.3 Hydrogen Production Capacity

The evolution of hydrogen production capacity underscores its pivotal role in the ongoing global energy transition. With a burgeoning demand for clean energy solutions, hydrogen emerges as a key player, particularly in sectors that are challenging to electrify directly, such as aviation and heavy industry. However, despite its potential, the pace of hydrogen's integration into the global energy landscape remains relatively slow compared to the targets set forth in the Paris Agreement. Projections indicate that by 2030, hydrogen could constitute a mere 0.5% of global energy consumption, with that figure potentially rising to 5% by 2050. Europe, in particular, is poised to lead in hydrogen adoption, with policies geared toward achieving 11% of final energy consumption from hydrogen by 2050. The outlook for hydrogen's future underscores its critical importance as a clean energy vector. Transitioning from fossil fuel-derived hydrogen to cleaner alternatives is imperative for combating climate change. According to IRENA, hydrogen and its derivatives would need to cover 14% of global final energy consumption by 2050 to limit global warming to 1.5°C. Currently, the majority of hydrogen production worldwide relies on fossil fuels, exacerbating climate change. Urgent action is required to shift toward clean hydrogen production methods, necessitating a fivefold increase in global hydrogen production by 2050, largely through renewable energy expansion and electrolyzer capacity enhancement.

Furthermore, the international trade of green hydrogen and its derivatives will play a crucial role in connecting low-cost production zones with high-demand areas. This will entail the development of new supply chains and the harmonization of regulatory frameworks to incentivize the green hydrogen market. These prospects underscore the significance of hydrogen, particularly green hydrogen, as a vital future energy vector for transitioning to a low-carbon economy. To realize its potential in reducing CO_2 emissions and meeting future energy needs sustainably, the development of favorable policies, reliable infrastructure, and international alignment of standards and regulations are paramount.

2.1.5.4 Main Sustainability Criteria

When considering the sustainability of hydrogen production methods, various critical factors must be carefully assessed to ensure that the chosen approach aligns with environmental, economic, and social objectives. First, the energy source utilized plays a pivotal role, as it determines the overall carbon footprint of hydrogen production. Methods relying on renewable energy sources like solar, wind, or hydroelectric power are generally regarded as more sustainable compared to those dependent on fossil fuels, which contribute significantly to GHG emissions.

Second, the amount of GHG emissions generated during the production process is a crucial consideration. Reducing emissions, particularly CO_2, is imperative for mitigating climate change. Therefore, methods that emit lower levels of GHGs or utilize

CCS technologies to sequester emissions are preferred, aligning with global decarbonization efforts. Water consumption is another vital criterion, especially in regions prone to water scarcity. Hydrogen production processes, such as electrolysis, can be water-intensive. Therefore, adopting methods that minimize water consumption or utilize nonpotable water sources can help alleviate pressure on freshwater resources and promote sustainable water management practices.

Furthermore, the efficiency of energy conversion into hydrogen is essential for optimizing resource utilization and minimizing costs. Higher efficiency translates to less energy waste and lower production costs, making the method more economically viable in the long run. Additionally, considering the environmental and biodiversity impact is crucial to safeguarding ecosystems and natural habitats. Methods that minimize pollution and habitat destruction are preferred, promoting ecological sustainability.

Lastly, the social acceptance and impact on local communities cannot be overlooked. Sustainable hydrogen production should contribute positively to communities by creating employment opportunities, respecting indigenous rights, and fostering equitable development. Engaging with stakeholders and addressing social concerns ensures that hydrogen projects are embraced by local populations, enhancing their long-term viability and contributing to broader societal well-being.

2.1.5.5 Infrastructure Requirements

Developing the infrastructure necessary to support the increased production and utilization of hydrogen is a multifaceted endeavor that requires careful consideration of technical, economic, and environmental factors. At the core of this effort is the establishment of production facilities capable of generating hydrogen through various methods, including electrolysis and reforming processes. These facilities must be strategically located to optimize access to renewable energy sources, such as solar and wind power, which are essential for producing green hydrogen with minimal carbon emissions. Furthermore, the development of distribution networks is crucial for transporting hydrogen from production sites to end users efficiently and safely. This involves the construction of pipelines, storage facilities, and transportation vessels capable of handling hydrogen in its gaseous or liquid form. Additionally, the adaptation of existing energy infrastructure, such as natural gas pipelines, may offer opportunities to repurpose existing assets for hydrogen transport, albeit with necessary modifications to ensure compatibility and safety.

Hydrogen storage presents another critical aspect of infrastructure development, particularly for managing the intermittency of renewable energy sources and fluctuating demand. Various storage technologies, including compressed gas tanks, liquid hydrogen tanks, and chemical storage systems, must be deployed to ensure a reliable supply of hydrogen for various applications. Moreover, the establishment of hydrogen refueling stations is essential for supporting the widespread adoption of hydrogen fuel cell vehicles, requiring careful planning and investment in infrastructure deployment.

Finally, standards, regulations, and training programs play a pivotal role in ensuring the safety, reliability, and interoperability of hydrogen infrastructure. Clear technical standards and regulatory frameworks must be established to govern the design, construction, operation, and maintenance of hydrogen facilities and equipment.

Additionally, comprehensive training programs are necessary to educate operators, emergency responders, and other stakeholders on best practices for handling hydrogen safely and effectively. Overall, the development of hydrogen infrastructure represents a complex yet essential endeavor in facilitating the transition to a sustainable, low-carbon energy future.

2.2 MAIN METHODS OF HYDROGEN PRODUCTION

2.2.1 STEAM REFORMING

Steam reforming, also known as SMR, is a process that converts hydrocarbons into a synthesis gas by reacting them with steam. This process, as shown in Figure 2.4, employs a nickel-based catalyst and occurs at high temperatures (between 840 and 950°C) and moderate pressure (between 20 and 30 bar). The resulting gas is a complex mixture comprising hydrogen, CO, CO_2, CH_4, water, and residual hydrocarbons ($H_2 + CO + CO_2 + CH_4 + H_2O$). While natural gas is the most commonly used feedstock for steam reforming, other hydrocarbons such as CH_4 or naphtha can also be utilized. The choice of hydrocarbon, also known as the feedstock for steam reforming, depends on the desired composition of the synthesis gas and the required hydrogen purity, leading to the adoption of different processes. Concerning the feedstock for steam reforming, light hydrocarbons such as natural gas, LPG, and naphtha, with

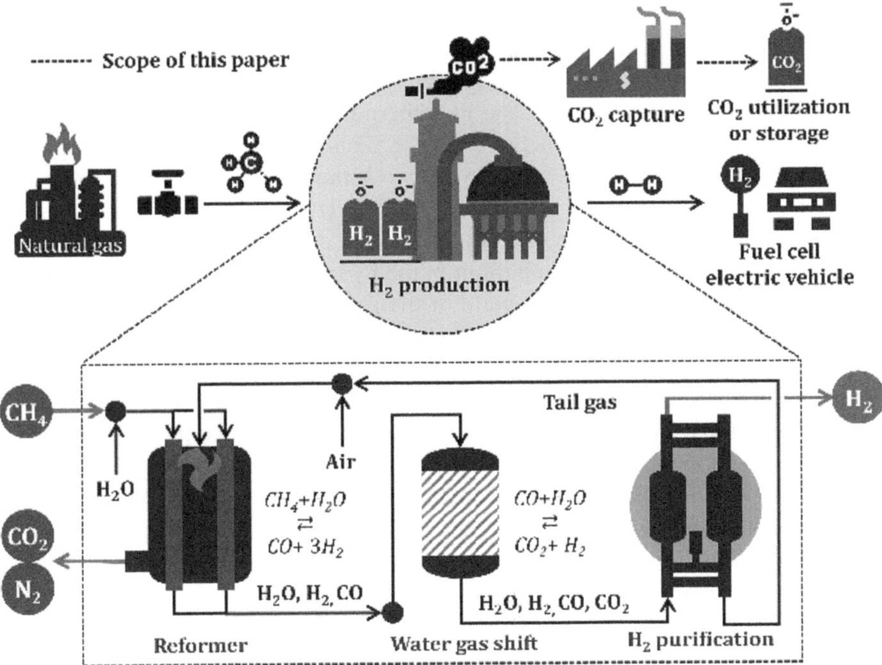

FIGURE 2.4 Schematic of industrial hydrogen production using steam methane reforming. (Adapted from ref. [5].)

boiling points not exceeding 200 to 220°C, are typically employed. Alcohols like methanol and ethanol are also options. The presence of sulfur in these feedstocks can inhibit the efficiency of the catalyst during steam reforming. Therefore, a pre-desulfurization step is necessary to reduce the sulfur content to less than 0.5 ppm by mass. This step involves reacting the hydrocarbon feedstock with hydrogen to produce hydrocarbons and hydrogen sulfide.

A subsequent step, the water–gas shift (WGS) reaction, increases the hydrogen concentration and allows for further hydrogen recovery. Although this process is well-established, it raises several technical and other questions, particularly in terms of efficiency, costs, environmental impact, and specific technical challenges.

2.2.1.1 Chemical Mechanism

Steam reforming involves the chemical reaction of hydrocarbons, typically CH_4, with steam (H_2O), in the presence of a high-temperature catalyst. Here's a detailed description of the chemical mechanism in several steps:

- Step 1: Primary Reforming Reaction

  The first step is the primary reforming reaction, where CH_4 reacts with steam to produce CO and hydrogen (H_2). This reaction is endothermic, meaning it absorbs heat.

$$CH_4(g) + H_2O(g) \rightarrow CO(g) + 3H_2(g)$$

  This chemical reaction shows that CH_4, in the presence of steam (H_2O), is converted into CO and hydrogen (H_2). It plays a fundamental role in industrial hydrogen production. The endothermic nature of the reaction means it requires an input of energy to occur. This energy is typically supplied in the form of heat, necessitating the reaction to occur at high temperatures, typically between 700°C and 1000°C. The required heat is often provided by the combustion of a portion of CH_4 or other available fossil fuels. This step is particularly important in the hydrogen production process as it defines the initial hydrogen yield of the process. The amount of hydrogen produced in this step directly affects the overall efficiency of the hydrogen production process. Optimizing this step involves managing the temperature, the steam-to-carbon ratio (H_2O/CH_4 ratio), and the choice of catalyst to maximize hydrogen production while minimizing energy consumption and the formation of undesirable by-products. The primary reforming step is essential in the hydrogen production process as it converts hydrocarbons into a mixture of CO and hydrogen, which can then be further processed to increase hydrogen purity. Effective management of this step is crucial for the efficiency and sustainability of hydrogen production.

- Step 2: Water–Gas Shift Reaction

  The WGS reaction step is a crucial phase in the steam reforming process for hydrogen production. This reaction increases the hydrogen yield while converting CO, an undesirable by-product, into CO_2, which can be more easily separated from the gas mixture. During this step, the CO produced in the first step can react with additional steam in a reaction called the WGS reaction, producing more hydrogen and CO_2. This reaction is exothermic, releasing heat.

$$CO(g) + H_2O(g) \rightarrow CO_2(g) + H_2(g)$$

Unlike the primary reforming reaction, the WGS reaction is exothermic, meaning it releases heat. This has significant implications for the thermal management of the overall hydrogen production process. The temperatures typically employed for the WGS range between 200°C and 400°C for high-temperature shift (HTS) and between 190°C and 250°C for low-temperature shift (LTS). This reaction increases the total amount of hydrogen produced by the process. As the reaction releases heat, it is important to effectively manage this energy, either by recovering it for other uses or by controlling the reactor temperature to optimize the reaction. Optimizing this step focuses on maximizing the conversion of CO to CO_2 to increase hydrogen production while minimizing the formation of undesirable by-products. This often involves the use of two reactors in series: one for HTS followed by one for LTS, to maximize conversion efficiency. The WGS reaction is a key step in increasing the hydrogen yield of the reforming process and adjusting the composition of the synthesis gas by reducing the CO concentration. Heat management and catalyst selection are crucial aspects for optimizing this reaction.

2.2.1.2 Kinetic Models

Kinetic models underpin the scientific understanding and optimization of reforming processes, particularly SMR, crucial for industrial hydrogen production. These models encapsulate a sophisticated network of equations delineating the kinetics of primary and secondary reactions, elucidating the intricacies of chemical transformation and rate determination. In SMR, for instance, the primary reforming step, where CH_4 reacts with steam to produce CO and hydrogen, is governed by a rate equation reliant on the partial pressures of CH_4 and water vapor, alongside the rate constant specific to SMR. Similarly, the WGS reaction, pivotal in adjusting hydrogen yield while converting CO to CO_2, is characterized by its own rate expression dependent on partial pressures, exemplifying the complexity of reaction kinetics within reforming processes.

Moreover, these models entail meticulous considerations such as the temperature-dependent nature of rate constants, typically described by the Arrhenius equation, and the dynamic variations in reactant partial pressures dictated by operational conditions and conversions. In-depth modeling may encompass additional parameters like catalyst deactivation kinetics and total pressure effects, crucial for accurate predictions. The integration of kinetic equations with comprehensive mass and energy balances facilitates predictive insights into reformer performance and aids in the optimization of operational parameters. Advanced computational tools, ranging from process simulation software to computational fluid dynamics (CFD) techniques, further refine model precision, enabling nuanced analyses of complex reforming systems essential for advancing hydrogen production technologies.

2.2.1.3 Thermodynamics of the Process

The hydrogen production process through reforming involves a series of intricate chemical reactions governed by fundamental thermodynamic principles, crucial for understanding the underlying dynamics of the system. Among these, SMR stands out as the predominant method, catalyzing the conversion of CH_4, typically sourced from natural gas, with steam in the presence of a nickel-based catalyst. This initial step, an endothermic reaction demanding elevated temperatures around 700–1100°C, serves as the cornerstone for hydrogen generation on an industrial scale. Subsequently, the WGS reaction plays a vital role, facilitated by the catalyst, in further refining the product mixture by converting CO into CO_2 and hydrogen (H_2) in either high-temperature or low-temperature stages, thereby enhancing hydrogen yield and purity.

In concert with the complex kinetics of reforming reactions, thermodynamic models provide a rigorous framework for elucidating the interplay of energy and entropy changes, as well as the feasibility and spontaneity of reactions under varying conditions. Gibbs free energy, a central concept in thermodynamics, serves as a barometer of reaction spontaneity, with its calculation involving enthalpy and entropy terms. The equilibrium constants, dictated by temperature, aptly reflect the thermodynamic favorability of reactions, encapsulating the dynamic equilibrium between reactants and products. Such rigorous thermodynamic analyses allow for the determination of optimal process parameters, such as temperature and pressure, essential for maximizing hydrogen production efficiency while minimizing undesirable by-product formation, thus informing the design and operation of reforming processes.

Practical implementation of thermodynamic insights into reforming processes necessitates a delicate balance between ideal thermodynamic conditions and real-world constraints, including kinetic limitations and material considerations. While thermodynamic models provide valuable insights into the theoretical limits of reaction outcomes, kinetic studies and economic analyses complement these by offering practical perspectives on process optimization. Integrating these multidisciplinary approaches facilitates the development of reforming processes that strike a harmonious balance between thermodynamic efficiency, technical feasibility, and economic viability, thereby advancing hydrogen production technologies toward sustainable and scalable solutions.

2.2.1.4 Types of Raw Material and Impact of Its Composition on the Process

The production of hydrogen through reforming methods, particularly steam reforming of hydrocarbons, involves a complex interplay of various raw materials that serve as substrates for the chemical reactions driving the process. Among these primary materials, hydrocarbons, predominantly sourced from natural gas, play a pivotal role as the primary feedstock for hydrogen generation. Hydrocarbons encompass a range of compounds, including CH_4, ethane, propane, and butane, each with distinct chemical properties that influence their reactivity and the efficiency of hydrogen production. Additionally, naphtha and heavier hydrocarbons can also be utilized, albeit less frequently, contributing to the diversity of feedstock options available for reforming processes. The selection of hydrocarbon feedstock is often dictated by factors such as availability, cost, and desired hydrogen yield, with natural gas emerging as the preferred choice due to its abundance and favorable reactivity.

Water serves as another indispensable raw material in reforming processes, functioning as a reactant in steam reforming reactions where it undergoes chemical transformation with hydrocarbons to produce hydrogen and carbon oxides. The steam reforming reaction, being highly endothermic, necessitates the input of significant thermal energy to drive the process forward. This requirement underscores the importance of precise control over process parameters, including temperature, pressure, and steam-to-carbon ratio, to optimize hydrogen production efficiency while mitigating energy consumption. Furthermore, the purity and quality of water used in the reforming process can influence reaction kinetics and catalyst performance, emphasizing the need for high-grade water sources and effective water treatment systems to maintain process integrity.

In certain reforming variants, such as autothermal reforming (ATR), the introduction of oxygen or air serves to facilitate the partial combustion of hydrocarbons within the reactor, generating the heat necessary to sustain reforming reactions. This controlled oxidation process offers advantages in terms of process simplicity and energy efficiency, as it obviates the need for external heat sources. However, precise control over oxygen supply is paramount to prevent complete combustion and ensure optimal hydrogen yield. Additionally, the selection of appropriate catalyst materials is instrumental in promoting desirable reaction pathways while mitigating undesirable side reactions, underscoring the intricate interplay between raw materials, reaction conditions, and catalytic activity in reforming processes.

2.2.1.5 Description of Main Equipment and Operational Conditions

The industrial production of hydrogen via SMR involves a series of meticulously controlled processes and equipment to ensure efficient and reliable production. Crucial components include a desulfurizer, which purifies natural gas by removing sulfur compounds that could hinder subsequent catalytic reactions. The primary reforming reactor serves as the central unit, where a mixture of natural gas and steam undergoes a high-temperature reaction over a catalyst, producing a blend of CO and hydrogen. Heat for this endothermic process is provided by a furnace, meticulously designed to maintain the requisite high temperatures without damaging equipment.

Another critical element is the WGS reactor, essential for converting CO into additional hydrogen and CO_2. Following reforming, gas mixtures often require purification to produce high-purity hydrogen. This is achieved through systems like Pressure Swing Adsorption (PSA), which selectively captures impurities at high pressure and releases them at low pressure, yielding pure hydrogen. Compressors then increase the pressure of the purified hydrogen for storage or transportation, an essential step in the process. Operational conditions play a fundamental role in optimizing the efficiency and efficacy of the SMR process. Maintaining temperatures between 700°C and 1000°C in the reforming reactor is critical for maximizing the endothermic reaction between CH_4 and steam. The WGS reaction typically occurs at temperatures ranging from 200°C to 400°C, often involving both high- and low-temperature phases to enhance CO conversion. Pressures within the reactor, typically between 3 and 25 bars, also impact hydrogen production rates, with higher pressures favoring production but potentially increasing costs.

Moreover, careful management of the steam-to-carbon ratio in the reforming reactor is essential to minimize solid carbon formation and optimize hydrogen yield. Nickel-based catalysts are commonly employed for their efficiency, but precise composition and preparation are crucial to their effectiveness and longevity. Additionally, rigorous sulfur compound reduction, by-products management, and safety protocols are integral to ensuring both operational efficiency and environmental sustainability throughout the hydrogen production process.

In Table 2.3, the summary of parameters affecting hydrogen production through steam reforming is presented. Key factors such as temperature, pressure, and steam-to-carbon ratio play critical roles in optimizing efficiency, cost-effectiveness, and

TABLE 2.3
Summary of Parameters Affecting Hydrogen Production through Steam Reforming

Parameter	Influence
Temperature	• Directly affects reaction equilibrium • Higher temperatures favor endothermic reactions like methane reforming, increasing hydrogen yield • Optimal temperature management minimizes carbon deposition and preserves catalyst performance • Higher temperatures require more energy, balancing efficiency with operational costs
Pressure	• Influences reaction equilibrium • Higher pressures favor the methane reforming reaction but can reduce hydrogen yield due to unfavorable equilibrium shifts • Accelerates reaction kinetics and separation efficiency but increases coking risk • High-pressure operations demand robust equipment, impacting construction and maintenance costs • Optimal pressure balancing efficiency, coking risk, and equipment costs is essential
Steam-to-Carbon Ratio (S/C)	• Critical for minimizing carbon deposition and maximizing hydrogen production • Too low ratios increase coking risk, while too high ratios elevate operational costs
Catalyst Composition	• Nickel-based catalysts are common but their activity, coking resistance, and durability depend on sulfur and other poisons
Feedstock Purity	• Impurities like sulfur and nitrogen reduce catalyst efficiency and must be removed via pretreatment
Dilution Rate with Other Gases	• Affects reaction chemistry and system heat capacity, influencing energy consumption and overall efficiency
Equipment Operating Conditions	• Influences overall process performance and reliability
By-product Management	• Effective management impacts environmental efficiency and hydrogen purity

hydrogen quality. Temperature control is essential for maximizing hydrogen yield while minimizing carbon deposition on catalysts. Pressure management balances reaction kinetics, separation efficiency, and equipment costs. Additionally, the steam-to-carbon ratio influences coking risk and operational expenses. The composition of catalysts, feedstock purity, dilution rates, equipment conditions, and by-product management are equally crucial for achieving sustainable, high-purity hydrogen production.

2.2.1.6 Catalysts: Their Role, Influence, and Types

Catalysts are indispensable in the steam reforming process due to their pivotal role in enhancing reaction kinetics, improving efficiency, and facilitating selectivity. First, catalysts lower the activation energy required for the reaction to proceed, thereby significantly increasing the reaction rate. This acceleration allows for a more rapid conversion of CH_4 and water into hydrogen and CO_2, essential components of the reforming process. Moreover, catalysts play a crucial role in controlling the temperature of the reaction, which is vital for maintaining process efficiency and safety, particularly given the highly endothermic nature of reforming reactions.

In steam reforming, catalysts typically consist of metals supported on inert materials, with nickel being the most widely used due to its efficiency and relatively low cost. Nickel catalysts, often supported on alumina, exhibit high activity and serve as the backbone for many industrial reforming processes. However, for applications demanding superior performance, platinum and rhodium from the platinum group metals are employed, albeit at a higher cost. These metals offer excellent catalytic activity and robust resistance to catalytic poisons, making them suitable for high-performance reforming applications. Furthermore, catalysts can be tailored by modifying or doping them with additional elements to optimize their performance and enhance their resistance to coking, a process where carbon accumulates on the catalyst surface and diminishes its efficacy. Ruthenium and iridium are also utilized in certain reforming processes, each offering unique advantages such as operating at relatively lower temperatures and providing high activity and stability, respectively. This intricate interplay between catalyst composition, structure, and function underscores the critical role of catalysts in driving the steam reforming process toward increased efficiency, selectivity, and sustainability. Given the above-mentioned explanations, the advantages and limitations of this process are summarized in Table 2.4.

2.2.2 Partial Oxidation

Partial oxidation is a key method for hydrogen production, especially in industries where access to large quantities of hydrogen is crucial for various chemical processes. This technique involves the chemical reaction between a hydrocarbon and a limited amount of oxygen, resulting in the production of hydrogen and CO. This method is particularly suitable for treating heavy hydrocarbons or those containing impurities that make their conversion by other methods less efficient. PO is a thermochemical conversion technology used to transform hydrocarbons into hydrogen and CO through combustion in the presence of an amount of oxygen lower than that

TABLE 2.4

Summary of Advantages and Limitations of Steam Reforming

Aspect	Advantages	Challenges
Efficiency	• High conversion efficiency of hydrocarbons into hydrogen	Energy-intensive process requiring high temperatures and pressures
Economies of scale	• Cost per unit of hydrogen decreases with increasing plant capacity	• Carbon emissions as a by-product, raising environmental concerns
Established technology	• Well-established with decades of operational experience	• Catalyst issues, such as coking and sintering at high temperatures
Infrastructure	• Utilizes existing natural gas infrastructure for transportation and distribution	• Carbon formation leading to blockages and increased maintenance costs
Feedstock flexibility	• Can process various hydrocarbons like natural gas, LPG, naphtha, or biogas	• Hydrogen separation from other gases, requiring costly and energy-consuming technologies
Carbon capture potential	• Compatible with carbon capture and storage technologies, enabling cleaner hydrogen production	• Integration of carbon capture technology raises questions about feasibility, efficiency, and costs
High-purity hydrogen	• Produces high-purity hydrogen suitable for various applications	• Reliance on fossil fuels ties hydrogen production to market fluctuations and perpetuates dependence on nonrenewable resources
Upgradability	• Facilities can be adapted or upgraded for enhanced efficiency, capacity, or integration of new technologies	• Safety concerns due to handling of highly flammable hydrogen gas, as well as high temperatures and pressures required for the process

required for complete combustion. This reaction occurs at high temperatures, typically between 800°C and 1,500°C, and under pressure. It can be catalyzed or non-catalyzed, depending on the type of hydrocarbon being processed and the specifics of the desired process. There are currently two major PO processes that differ in the concrete implementation of the reactions: the Shell process and the Texaco process. The heat treatment method is also different: heat recovery by contact for Shell and quench cooling for Texaco.

2.2.2.1 Partial Oxidation Process and Its Applications

The PO process, a fundamental method in industrial hydrogen production, commences with the meticulous blending of a hydrocarbon feedstock, such as natural gas or heavier petroleum derivatives, with either oxygen or oxygen-enriched air. This meticulously engineered mixture is then carefully introduced into a specialized oxidation reactor. Within this reactor, the mixture undergoes a transformative journey under stringent conditions of high temperature, typically ranging between 1200 and 1500°C, and elevated pressure, often spanning 20 to 90 bar. These conditions are meticulously controlled to optimize the reaction kinetics and product yields. Catalysts are frequently employed within the reactor to enhance reaction rates and efficiency. The core of the process lies in the PO reaction, a highly exothermic process wherein the hydrocarbon molecules interact with oxygen to yield a synthesis gas rich in hydrogen (H_2) and CO.

The principal reaction, exemplified here for CH_4, a predominant component of natural gas, can be represented as follows:

$$CH_4 + 1/2\ O_2 \rightarrow CO + 2H_2$$

However, alongside this primary reaction, a suite of secondary reactions ensues, influenced by factors such as temperature, pressure, and the composition of the reaction mixture. These secondary reactions may lead to the formation of by-products such as CO_2 and water (H_2O). Moreover, the intricacies of the PO process extend beyond the primary reactor, as downstream stages witness further transformations. At exceedingly high temperatures, secondary reactions like hydrocracking, steam carbon gasification, reforming, the WGS reaction, and sulfur conversion manifest, further enriching the spectrum of products and intermediates, contributing to the complexity and versatility of the process.

Hydrogen generated through PO serves as a crucial raw material in several industrial sectors due to its versatility and reactivity. In the synthesis of ammonia, a pivotal process in fertilizer production, hydrogen plays a central role in reacting with atmospheric nitrogen to form ammonia via the Haber–Bosch process. This reaction, facilitated by iron-based catalysts, enables the efficient conversion of nitrogen into ammonia, which is essential for global agricultural productivity. Additionally, hydrogen is integral to the production of methanol through the methanol synthesis reaction, where it reacts with CO to yield methanol, a fundamental precursor in the synthesis of various chemicals and fuels. This process is typically catalyzed by metal-based catalysts under high-pressure conditions, enabling the synthesis of methanol on an industrial scale.

Furthermore, hydrogen finds extensive application in petroleum refining processes, particularly in hydro-treating heavy fractions of crude oil. In this context, hydrogen acts as a reagent in various hydro-treatment reactions, such as hydrodesulfurization and hydrocracking, aimed at reducing the sulfur content and upgrading the quality of petroleum products. Hydrogen's ability to react with sulfur and other impurities present in crude oil fractions results in the production of cleaner fuels with improved environmental performance. Consequently, the utilization of hydrogen in petroleum refining not only enhances the efficiency of refining processes but also contributes to the production of cleaner and higher-quality fuels, aligning with global sustainability goals.

Given the above-mentioned explanations, the advantages and limitations of this process are summarized in Table 2.5.

2.2.3 Water Electrolysis

Electrolysis, a process utilizing electricity, divides water into its constituent elements, hydrogen and oxygen. Two primary methods are employed: alkaline electrolysis (AE), utilizing potassium hydroxide (KOH) or sodium hydroxide (NaOH) as the electrolyte, and Proton Exchange Membrane (PEM) electrolysis, employing

TABLE 2.5

Summary of Advantages and Limitations of Partial Oxidation

Advantages	Challenges
• Flexibility in the types of hydrocarbons processed, enabling the use of cheaper or locally available feedstocks	• Need to manage high temperatures and extreme reaction conditions, which can pose technical challenges and higher investment costs
• Production of hydrogen at high temperatures, which can be directly utilized for certain industrial processes without the need for additional reheating	• Management of undesirable by-products such as CO2 and NOx, which can have significant environmental impacts
• Ability to handle feedstocks with high levels of impurities	• Process optimization to enhance efficiency and reduce operational costs, particularly concerning oxygen consumption and hydrogen yield

a polymer membrane selectively permeable to protons. In AE, an electric current breaks down water molecules into hydrogen and oxygen gases, while in PEM electrolysis, the membrane facilitates the separation of hydrogen and oxygen under the influence of an electric field. Each method has distinct advantages and applications, with AE commonly utilized in industrial contexts and PEM electrolysis gaining attention for smaller-scale and renewable energy-integrated setups.

2.2.3.1　Principles and Techniques of Electrolysis

Electrolysis of water is a key method for hydrogen production, especially when powered by renewable energy sources. With the growing need to reduce GHG emissions, hydrogen produced through water electrolysis offers a clean alternative to fossil fuels. Hydrogen can serve as an energy carrier, enabling the storage and transportation of renewable energy. Water electrolysis is particularly compelling when coupled with renewable energy sources such as wind, solar, or hydroelectric power, enabling hydrogen production without direct CO_2 emissions, crucial for carbon neutrality goals. Technological advancements in electrolyzers have improved their efficiency, durability, and reduced costs. Types of electrolyzers include AE, PEM electrolysis, and Solid Oxide Electrolysis Cell (SOEC), each with its own advantages and applications. The cost of hydrogen produced by electrolysis largely depends on the cost of electricity used. Low prices of renewable electricity can thus make this method more competitive compared to hydrogen production from fossil fuels (such as CH_4 reforming). Many countries have implemented policies to support the development of green hydrogen, including subsidies, tax incentives, and national goals to increase hydrogen utilization in various sectors like transportation, industry, and heating. Despite its advantages, water electrolysis faces challenges such as high initial investments and managing the demand and supply of renewable electricity for continuous production. However, hydrogen's potential to facilitate the decarbonization of hard-to-electrify sectors offers significant opportunities for its future development. These factors contribute to shaping the context in which water electrolysis for hydrogen production exists, making it an essential component of global energy transition strategies.

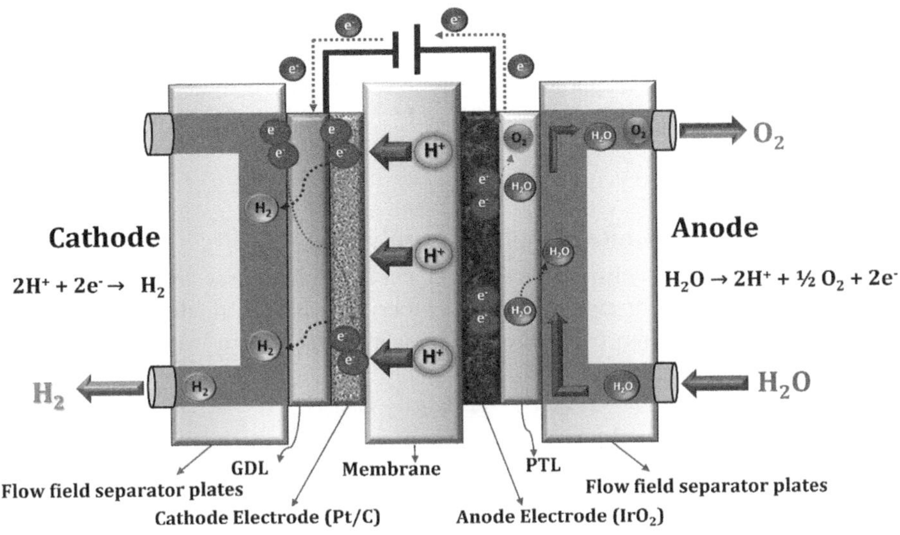

FIGURE 2.5 A schematic for standard electrolysis.

(Adapted from ref. [6].)

Water electrolysis is a chemical process that utilizes electricity to decompose water (H_2O) into its elemental components, hydrogen (H_2) and oxygen (O_2). This process takes place within a device called an electrolyzer (Figure 2.5). A typical electrolyzer consists of two electrodes (an anode and a cathode) submerged in an electrolytic solution, which can be either pure water or water with added electrolytes to enhance conductivity.

A continuous electric current is applied between the two electrodes. The electricity must be sufficiently powerful to overcome the water electrolysis voltage, which is approximately 1.23 volts at 25°C but varies with temperature and pressure. At the cathode (negative electrode), water is reduced to form gaseous hydrogen and hydroxide ions (OH^-) according to the reaction:

$$2H_2O + 2e^- \rightarrow H_2 + 2OH^-$$

At the anode (positive electrode), water is oxidized to release gaseous oxygen and hydrogen ions (H^+), as per the reaction:

$$2H_2O \rightarrow O_2 + 4H^+ + 4e^-$$

The collection of gases, hydrogen at the cathode and oxygen at the anode, can be done separately or together, contingent upon the design of the electrolyzer. Various types of electrolyzers exist, each possessing distinct characteristics. AE utilizes alkaline solutions like KOH or NaOH as the electrolyte. PEM electrolysis employs a solid membrane selectively allowing protons to pass through, thereby ensuring higher hydrogen purity and enhanced energy efficiency. SOEC functions at elevated temperatures and employs specialized ceramics as electrolytes, resulting in superior thermal efficiency. This technology of water electrolysis is pivotal for the production of green hydrogen, particularly when driven by renewable energy sources, thus playing a significant role in mitigating carbon emissions and fostering the transition toward sustainable energy systems.

2.2.3.2 Thermodynamics of Water Electrolysis

Water electrolysis for hydrogen production is a fascinating process that can be analyzed and understood through principles of thermodynamics, mathematical models, and electrolyzer efficiency calculations. The overall reaction for water electrolysis is:

$$2H_2O(l) \rightarrow 2H_2(g) + O_2(g)$$

This reaction requires energy input to occur, expressed by the Gibbs free energy (ΔG). At 25°C (298 K), ΔG for this reaction is +237.13 kJ/mol, indicating it is nonspontaneous and requires external energy input. The standard electrolysis potential, $\Delta G = -nFE$, where n is the number of moles of electrons exchanged (4 moles of electrons for the overall reaction), F is the Faraday constant (96485 C/mol), and E is the electrochemical potential. The standard electrolysis potential is approximately 1.23 volts. The enthalpy change of the reaction, ΔH, is +285.83 kJ/mol at 25°C, indicating it is endothermic and absorbs heat from its surroundings.

The cell voltage required for electrolysis is higher than the theoretical potential due to ohmic losses, concentration losses, and activation losses. The simplified model is given by:

$$E = E_0 + \eta_a + \eta_c + i_R$$

where E_0 is the theoretical voltage, η_a and η_c are the overpotentials at the anode and cathode, respectively, i is the current, and R is the internal resistance of the cell.

Energy efficiency can be calculated as the ratio of the useful energy obtained (the chemical energy of hydrogen) to the electrical energy consumed.

$$\eta_{energy} = \left(\left(\text{Calorific value of } H_2\right) / \left(\text{Electrical energy consumed}\right)\right) \times 100\%$$

Faradaic efficiency measures the efficiency with which electrons are used to produce hydrogen and oxygen.

$$\eta_{faradaic} = ((\text{Theoretical amount of } H_2 \times \text{produced}) / (\text{Actual amount of} \times H_2 \text{ produced from the applied current})) \times 100\%$$

Overall efficiency can be expressed as the product of energy efficiency and Faradaic efficiency.

$$\eta_{global} = \eta_{energy} \times \eta_{faradaic}$$

These models and calculations are crucial for the design, optimization, and comparison of electrolyzers in terms of cost, efficiency, and economic viability for industrial and energy applications.

2.2.3.3 Different Types of Electrolysis Technologies

There are several types of electrolysis technologies used to produce hydrogen from water, each with its own characteristics and applications. Here are the main types of electrolyzers:

i. **Alkaline Electrolysis (AE)**: This is the most mature and widely used technology. It operates by passing an electric current through water containing an alkaline electrolyte solution (often a solution of KOH or NaOH). Electricity breaks down water into oxygen and hydrogen. This method is known for its reliability and efficiency, although it requires relatively high currents and robust electrodes.

ii. **Proton Exchange Membrane (PEM) Electrolysis**: This technology uses a solid polymer membrane that serves both as an electrolyte and a gas separator. PEM electrolyzers can operate at higher current densities than AEs, allowing for faster hydrogen production from a smaller unit. They are also capable of starting and stopping quickly, making them suitable for applications that require rapid modulation based on electricity demand or supply (such as those related to intermittent renewable energies).

iii. **Solid Oxide Electrolysis (SOE)**: Solid oxide electrolyzers use solid ceramic as an electrolyte that conducts oxygen ions at high temperatures (typically between 700°C and 800°C). This technology is under development and promises superior energy efficiency by harnessing heat to reduce the electrical energy required for water decomposition. SOEs are particularly interesting for coupling with industrial heat sources or processes that generate excess heat.

iv. **Anion Exchange Membrane (AEM) Electrolysis**: A newer technology, AEM electrolyzers use a membrane that allows the passage of anions (mainly hydroxides). These systems combine some advantages of PEM and alkaline technologies, such as the potential to operate at low temperatures with less-corrosive electrolytes than those used in AE.

Each type of electrolyzer has its specific advantages and disadvantages in terms of cost, efficiency, durability, maintenance, and ability to integrate with different energy sources, including renewables. The choice of technology often depends on the specific application, energy availability, and economic and environmental goals. Table 2.6 shows the advantages and disadvantages of the different electrolysis technologies. The choice of technology often depends on the specific application, energy availability, and economic and environmental goals.

2.2.3.4 Factors Influencing the Efficiency of Electrolysis

The efficiency of electrolysis, which transforms electricity into hydrogen, is subject to various influential factors. Temperature stands out as a significant determinant, with higher temperatures typically enhancing electrolyte conductivity and reducing

TABLE 2.6

Summary of the Advantages and Limitations of Different Electrolyzers

Technology	Advantages	Disadvantages
Alkaline Electrolysis (AE)	• High technological maturity • Robustness and reliability • Reduced material costs	• Low current density • Limited dynamic response • Need for additional purification
PEM Electrolysis	• High current density • Rapid dynamic response • High-purity hydrogen	• High material cost • Sensitivity to impurities • Membrane durability issues
Solid Oxide Electrolysis (SOE)	• High energy efficiency • Integration with residual heat • Useful for certain industrial applications	• High development costs • High temperature requirements • Operational complexity
AEM Electrolysis	• Potential for reduced cost • Operates at low temperature • Less corrosive	• Limited technological maturity • Durability issues • Variable efficiency

overpotential, thereby boosting overall efficiency. However, elevated temperatures may accelerate the degradation of essential electrolyzer components. Pressure also plays a pivotal role, as operating at higher pressures can mitigate losses associated with hydrogen compression postproduction, albeit at the potential expense of increased equipment costs. Moreover, the concentration and type of electrolyte are critical, affecting both ionic conductivity and electrical resistance within the cell. Balancing these factors optimally maximizes efficiency while minimizing corrosion risks and material expenses.

Additionally, current and current density influence hydrogen production rates, with higher currents yielding increased production but potentially at the cost of greater ohmic losses and overpotential. Water purity is another essential consideration, as impurities can lead to electrode fouling or membrane damage, compromising both efficiency and durability. Cell design, encompassing electrode layout, configuration, and spacing, significantly impacts internal resistance and leakage current losses, further affecting overall efficiency. Moreover, the choice of catalysts, responsible for reducing overpotential, is crucial, although more efficient catalysts often come with higher costs. Finally, integrating electrolysis units with renewable energy sources can optimize energy utilization and reduce costs, ultimately enhancing efficiency and contributing to economic and environmental sustainability.

2.2.3.5 Impact of Innovations in Electrode Materials on Electrolysis Efficiency

Advancements in electrode materials wield significant influence over the efficiency of electrolysis, a pivotal process in hydrogen production. These advancements manifest in several critical aspects. First, superior electrodes can curtail electrochemical overpotential, reducing the energy threshold required to initiate water electrolysis reactions. By diminishing the energy demand to reach electrolysis potential, these materials directly bolster overall energy conversion efficiency. Second, heightened conductivity in advanced materials mitigates resistive losses within electrodes, diminishing internal resistance and minimizing energy losses in the form of heat. This enhancement enhances the process's efficiency by ensuring more effective electron transport.

Moreover, advanced materials offer enhanced durability and chemical stability, crucial attributes in harsh electrolytic environments characterized by high acidity, alkalinity, or temperature. Their resistance to corrosion and wear minimizes the need for frequent replacements and maintenance, thereby reducing operational interruptions and costs. Additionally, these materials often exhibit heightened catalytic properties, accelerating reactions at the electrodes while demanding less energy input. Novel catalysts, specially tailored to enhance reaction rates without relying on expensive metals like platinum, render electrolysis more cost-effective.

Furthermore, the development of materials featuring nanostructured or porous configurations amplifies the active surface area of electrodes, facilitating more efficient interaction with water molecules and, consequently, augmenting hydrogen production rates and overall efficiency. Beyond this, materials optimized for thermal management contribute to stable and efficient operation, particularly pertinent in HTE. Lastly, the utilization of less expensive or more abundant materials for electrodes substantially reduces manufacturing costs, heightening the competitiveness of hydrogen produced via electrolysis against other energy sources. In summation, the evolution of electrode materials is instrumental in crafting electrolyzers that are not only more efficient and durable but also economically viable, thereby propelling electrolysis as a compelling avenue for large-scale hydrogen production in support of cleaner energy transitions.

2.2.3.6 Environmental and Economic Challenges

The environmental impact of hydrogen produced through electrolysis hinges on various factors, notably the energy source powering the electrolyzer, the efficiency of the electrolysis technology, and the associated production and distribution processes. If renewable sources like wind, solar, or hydroelectric power drive the electrolysis process, the resulting hydrogen—termed "green hydrogen"—boasts a minimal carbon footprint. Conversely, reliance on fossil fuel-generated electricity significantly escalates hydrogen's carbon footprint. Electrolysis technologies differ in efficiency, affecting the environmental impact of hydrogen production. Higher efficiency technologies require less electricity for hydrogen production, curbing indirect emissions linked to electricity generation.

Furthermore, the manufacture of electrolyzers, hydrogen transport, and infrastructure maintenance carry environmental implications. Material production for electrolyzers, especially rare or precious metals used in some cell types, entails significant environmental impacts like mining and chemical processing. While electrolysis itself produces no CO_2 or atmospheric pollutants during operation, emissions tied to the energy source must be considered. Additionally, industrial processes for electrolyzer components may emit pollutants. The water demand for electrolysis is another environmental factor; although water is plentiful, electrolysis often necessitates large quantities of purified water, posing challenges in regions with limited water resources. Material durability and recyclability in electrolyzers also contribute to the environmental footprint, with efforts to develop more durable and recyclable materials crucial for minimizing impact.

2.2.3.7 The Cost of Hydrogen Production

Hydrogen production costs vary considerably depending on the method used, input costs, technology, and production scale. Electrolysis relies on electricity, with costs influenced by electricity prices and electrolyzer efficiency, ranging between $2 and $7 per kilogram. SMR, due to the affordability of natural gas, costs between $1 and $3 per kilogram but generates CO_2 emissions, potentially incurring additional costs for CCS technologies. Coal gasification costs usually range between $2 and $4 per kilogram, with additional expenses for emissions treatment. These costs are influenced by various regional and global factors, including energy policies, carbon pricing mechanisms, renewable energy subsidies, and technological advancements. As the world shifts toward a low-carbon economy, the relative costs of hydrogen production methods may evolve, particularly with charges for GHG emissions or declining renewable technology costs.

Hydrogen production costs vary by region due to differences in energy prices, infrastructure, technology adoption, and policy support. In Europe, costs are influenced by strong commitments to renewable energy and stringent environmental regulations, with electrolysis costs between $3 and $7 per kilogram and SMR with CCS costs from $1.5 to $3 per kilogram. In Asia, Japan faces higher costs due to reliance on imported energy, with electrolysis costs from $4 to $8 per kilogram. China benefits from lower production costs, particularly for hydrogen derived from coal gasification at $1 to $2 per kilogram, but faces environmental drawbacks. South Korea's focus on green hydrogen results in electrolysis costs from $3 to $7 per kilogram.

In North America, the United States and Canada benefit from abundant natural gas resources and advanced technology, with SMR costs ranging from $1 to $2.5 per kilogram and electrolysis costs from $2.5 to $6 per kilogram. Both countries focus on blue hydrogen, produced using SMR with CCS, leveraging natural gas reserves and strong environmental policies.

Technological advancements, economies of scale, and policy support are expected to reduce hydrogen production costs across all regions, particularly for green hydrogen, making it more competitive globally. This comparison provides a comprehensive overview of the economic feasibility of hydrogen production in different markets, highlighting regional disparities and the potential for future cost reductions. Figure 2.6 illustrates the hydrogen production costs in the USA.

From another point of view, the cost of hydrogen production varies significantly depending on the method employed, with electrolysis being one of the prominent techniques. Electrolysis costs can range from $3 to $8 per kilogram of hydrogen when using renewable energy sources, while conventional energy sources can raise the cost to $6 to $9 per kilogram. In contrast, SMR, the conventional method, typically costs $1 to $3 per kilogram. Despite its higher costs, electrolysis offers the advantage of producing "green hydrogen" when powered by renewables, significantly reducing carbon footprints. Coal gasification and CH_4 pyrolysis also present

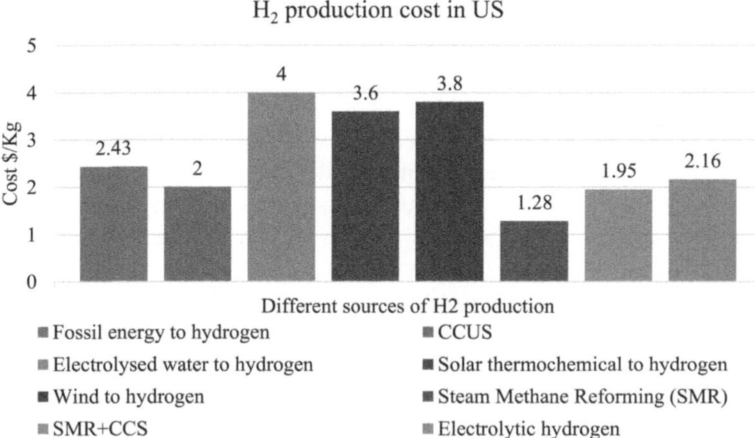

FIGURE 2.6 Hydrogen production cost in the USA (in $).

(Adapted from ref. [7].)

TABLE 2.7
Comparative Costs of Hydrogen Production Methods

Production Method	Cost (USD/kg of H_2)	Comments
SMR	$1–$3	Cheaper but emits CO_2, heavily dependent on natural gas prices
Electrolysis (Conventional Energy)	$6–$9	More expensive due to higher electricity prices from nonrenewable sources
Electrolysis (Renewable Energy)	$3–$8	Cost varies depending on renewable energy sources and production scale
Methane Pyrolysis	$2–$5	Less CO_2 emissions compared to SMR, but less mature technology and potentially costly
Coal Gasification	$2–$4	Relatively economical but highly polluting, dependent on coal prices and regulations

cost-effective alternatives, but their environmental impacts, especially in terms of CO_2 emissions, are higher. The choice of production method depends on factors such as energy availability, environmental goals, and economic considerations. Table 2.7 is a comparative analysis showing the approximate costs of hydrogen production by different methods.

2.2.3.8 Impact of Renewable Energy Integration on the Viability of Water Electrolysis

The integration of renewable energies into water electrolysis profoundly impacts both the economic and environmental viabilities of hydrogen production. One crucial aspect is the reduction in energy costs associated with electrolysis, as significant amounts of electricity are required. Utilizing renewable sources like wind, solar, or hydroelectric power can substantially lower these energy costs. These renewable sources offer low marginal costs once installations are in place, thus providing a cost-effective energy supply and improving the overall economics of electrolysis. Moreover, the integration of renewable energies allows for a reduction in the carbon footprint of hydrogen production. When renewable energies power the electrolysis process, the resulting hydrogen is often termed "green" hydrogen, indicating that it is produced without direct CO_2 emissions. This makes green hydrogen an attractive option for industries and transportation sectors striving to align with global climate objectives by reducing their environmental impact.

Furthermore, integrating electrolysis into energy grids enhances grid flexibility, serving as a means of load management. Excess electricity generated during periods of high renewable energy production can be utilized for electrolysis, stabilizing the grid by absorbing production peaks. This helps avoid grid overload issues and reduces the need for temporary shutdowns of renewable production, contributing to a more stable and reliable energy supply.

2.2.3.9　Security and Logistics

During hydrogen production via electrolysis, strict adherence to safety standards is paramount to mitigate the risks of fire, explosion, and chemical exposure. These standards encompass various facets of facility design, operation, and maintenance. Key safety measures include ensuring equipment complies with international standards such as ISO or IEC, dividing installations into zones based on explosion risk, implementing effective ventilation systems to prevent hydrogen gas accumulation, and installing gas detectors for continuous leak monitoring. Additionally, routine inspections and maintenance checks are essential to uphold equipment integrity, while personnel must undergo comprehensive training in safety procedures and emergency response protocols.

Moreover, it's imperative to have evacuation plans in place, accessible fire extinguishers for different fire classes, and proper handling and disposal procedures for chemical waste to prevent environmental contamination. Strict control of emissions into the atmosphere, including gases from electrolysis, is necessary to meet environmental standards. Obtaining relevant certifications like ISO 45001 for occupational health and safety management systems is advisable, coupled with regular audits to ensure ongoing compliance with safety regulations. By adhering to these stringent safety standards and protocols, the inherent risks associated with hydrogen production through electrolysis can be substantially mitigated, ensuring a safe working environment and regulatory compliance.

The amount of energy required to produce one kilogram of hydrogen by electrolysis depends on several factors, including the efficiency of the electrolysis technology used and operating conditions. However, this quantity can be estimated based on basic thermodynamic and electrochemical principles. Theoretical Calculation of Required Energy: hydrogen production by electrolysis involves the decomposition of water (H_2O) into oxygen (O_2) and hydrogen (H_2) according to the following reaction: $2H_2O(l) \rightarrow 2H_2(g) + O_2(g)$. For each mole of oxygen produced, two moles of hydrogen are produced. The theoretical potential required to decompose water is approximately 1.23 volts at standard temperature and pressure. However, in practice, overvoltages and internal resistances of the electrolyzer increase this value.

Practical Energy Calculation: in practice, electrolysis may require potentials of approximately 1.8 to 2.6 volts per cell due to internal losses and inefficiencies. Considering the overall efficiency of the system, the amount of electrical energy required to produce one kilogram of hydrogen is generally estimated as follows: electrical energy required is approximately 50 to 55 kWh/kg of hydrogen for modern high-efficiency electrolysis systems, ranging up to 65 kWh/kg or more for less efficient or older systems. These values take into account the energy needed to decompose water, as well as the additional energy required to overcome system efficiency losses, such as heat loss and electrical resistances.

2.2.4 Hydrogen Production from Nuclear energy

Nuclear energy can produce hydrogen through two main methods: thermochemical water splitting and HTE. These methods offer a low-carbon solution, leveraging the high temperatures provided by nuclear reactors to enhance the efficiency and sustainability of hydrogen production.

2.2.4.1 Thermomechanical Water Splitting

Thermochemical water splitting involves utilizing high temperatures to catalyze the decomposition of water (H_2O) into hydrogen (H_2) and oxygen (O_2). This process uses a series of chemical reactions, with cycles such as the iodine–sulfur (IS) cycle, the iron–chlorine cycle, and the copper–chlorine cycle (Cu–Cl) being particularly noteworthy. For example, the IS cycle operates at temperatures up to 850°C, making it ideal for integration with high-temperature nuclear reactors. The basic principle of thermochemical water splitting involves utilizing heat to drive a series of chemical reactions that ultimately decompose water into hydrogen and oxygen. These reactions occur cyclically, with intermediates being regenerated and reused within the cycle. The necessary high temperatures for these reactions range from 500°C to 2000°C, depending on the specific thermochemical cycle employed. Notable cycles include the IS cycle, which operates at moderate temperatures with high efficiency, the Cu–Cl cycle, which operates at even lower temperatures and is easier to integrate with nuclear reactors, and the IS cycle, which is similar to the IS cycle but involves different intermediate steps.

Nuclear reactors, particularly high-temperature gas-cooled reactors (HTGRs) and molten salt reactors (MSRs), are well-suited to provide the heat required for thermochemical water splitting. These reactors can achieve the high temperatures necessary for processes like the IS cycle, with HTGRs reaching up to 1000°C. The heat from the reactor is transferred to the thermochemical process through a heat exchanger. MSRs also achieve high temperatures and offer excellent heat transfer properties, making them suitable for various high-temperature applications, including thermochemical water splitting.

Thermochemical water splitting integrated with high-temperature nuclear reactors offers significant advantages, such as high efficiencies and zero CO_2 emissions since nuclear energy is carbon-free. This integration makes large-scale hydrogen production feasible, as nuclear reactors can efficiently utilize their excess heat in thermochemical cycles. Despite these benefits, the technology faces challenges, including the need for materials that can withstand high temperatures and corrosive environments. Ongoing research aims to optimize these cycles for greater efficiency, lower operational costs, and effective integration with existing nuclear reactors, paving the way for sustainable and industrial-scale hydrogen production.

2.2.4.2　High-temperature Electrolysis (HTE)

HTE using SOECs is another advanced hydrogen production method. Operating at temperatures between 700°C and 800°C, HTE enhances the electrolysis efficiency by reducing the required overpotential, thereby lowering the overall energy consumption. SOECs consist of an anode, a cathode, and a solid electrolyte, typically made of yttrium-stabilized zirconium oxide (YSZ), which conducts oxygen ions at high temperatures. The heat required for this process can be efficiently supplied by nuclear reactors, significantly improving the overall efficiency of hydrogen production. Integrating these methods into hydrogen production strategies not only supports the transition to a low-carbon economy but also maximizes the utilization of nuclear energy for sustainable hydrogen generation. HTE using SOECs is an advanced method for producing hydrogen by electrolyzing water at elevated temperatures, typically between 700°C and 800°C. This high-temperature environment significantly enhances the efficiency of the electrolysis process by reducing the electrical energy needed to split water molecules into hydrogen and oxygen. In SOECs, water fed to the cathode undergoes a reaction to produce hydrogen and oxide ions. These oxide ions then migrate through the solid oxide electrolyte to the anode, where they are oxidized to form oxygen gas. The process typically utilizes YSZ as the electrolyte, with nickel-based cermet for the cathode and lanthanum strontium manganite (LSM) for the anode.

Nuclear reactors, particularly HTGRs and MSRs, are ideal for providing the necessary heat for SOEC-based HTE. These reactors can operate at temperatures up to 1000°C, supplying both the heat and electricity required for efficient electrolysis. The integration of nuclear reactors with electrolysis units through heat exchangers ensures that the operational temperature remains optimal for SOECs, resulting in very high efficiencies for hydrogen production. This synergy between nuclear heat and electricity supply makes the overall system highly efficient, surpassing the efficiencies of conventional electrolysis methods.

HTE using SOECs offers several advantages, including efficiencies of up to 90% due to the high operating temperatures, which reduce the electrical energy required for electrolysis. The excess heat produced by nuclear reactors can be effectively utilized in the electrolysis process, further enhancing overall energy efficiency and sustainability. Additionally, the integration with nuclear energy results in low carbon emissions, as nuclear reactors provide a carbon-free source of electricity and heat. This makes HTE a significant contributor to a sustainable and low-carbon energy future, aligning with global decarbonization goals. Moreover, the scalability of HTE, facilitated by large-scale nuclear reactors, makes it feasible for industrial-scale hydrogen production, crucial for meeting the growing demand for hydrogen in various applications and supporting the transition to cleaner energy sources.

2.2.5　Hydrogen Production from Biomass

In this process, biomass (such as agricultural waste, forest residues, etc.) is converted into a gaseous mixture of CO, CO_2, and hydrogen using steam and/or oxygen as gasifying agents. The hydrogen is then separated from the other gases.

The use of biomass for hydrogen production stands as a promising frontier in the global endeavor to foster renewable energy sources and curtail GHG emissions. Biomass, encompassing organic materials like wood, agricultural residues, and organic waste, emerges as a renewable energy reservoir. Leveraging biomass for hydrogen synthesis offers the potential to diminish reliance on fossil fuels, marking a pivotal step toward sustainability. Diverse methods, including gasification, pyrolysis, and anaerobic digestion, enable the conversion of biomass into hydrogen, each method carrying distinct advantages, drawbacks, and technological requisites. Gasification, for instance, entails the thermochemical transformation of biomass into syngas, ripe with CO and hydrogen, with subsequent processing elevating hydrogen concentration. Conversely, pyrolysis engenders thermal decomposition of oxygen, yielding biochar alongside hydrogen, while anaerobic digestion fosters biogas—predominantly CH_4 and CO_2—subject to further refinement into hydrogen through steam reforming. Such endeavors hold promise not only for sustainable waste management but also for GHG emission mitigation, effectively closing the carbon cycle by rendering produced energy carbon-neutral, contingent upon biomass sourcing and process nuances. Despite the considerable benefits, challenges loom, encompassing energy efficiency, technology costs, and by-product management, necessitating substantial research and development investments. The advancement and adoption of hydrogen production technology from biomass hinge heavily on government policies, incentives, and financial backing, vital in fostering renewable energy adoption and fulfilling climate objectives. In sum, biomass-derived hydrogen presents a compelling avenue for bolstering the renewable energy segment within the global energy landscape while bolstering endeavors to mitigate climate change.

The diverse array of biomass types available for hydrogen production encompasses agricultural waste, forest residues, municipal solid waste, energy crops, organic industrial waste, and fats/oils. Ranging from crop residues like straw and rice husks to forestry by-products such as wood chips, these materials offer abundant and often underutilized resources ideal for energy conversion. Additionally, organic waste from urban areas, including food scraps, can be harnessed alongside specially cultivated energy crops like miscanthus and switchgrass. Furthermore, industrial by-products like sugarcane bagasse and brewery residues present valuable sources of energy potential. Notably, fats and oils, whether from used oils or animal fats, hold promise for hydrogen synthesis owing to their high energy content. However, the selection of biomass hinges on factors such as local availability, transportation costs, processing efficiency, and environmental sustainability, demanding careful assessment to optimize both economic viability and environmental impact.

2.2.5.1 Basic Principle of Biomass Conversion to Hydrogen

The fundamental concept of converting biomass into hydrogen revolves around various processes aimed at extracting hydrogen-rich gases from the organic constituents of biomass. These processes fall into three main categories: thermochemical, biochemical, and physicochemical conversions. In thermochemical conversion, methods like gasification involve subjecting biomass to high temperatures in the presence of a limited oxidizing agent, yielding a gas mixture known as syngas, rich in CO and hydrogen. Alternatively, pyrolysis entails heating biomass in the absence of oxygen, leading to the decomposition of organic matter into bio-oils, char, and gases, including hydrogen. Each thermochemical method offers unique advantages and challenges in terms of efficiency and hydrogen yield.

Biochemical conversion methods, on the other hand, harness microorganisms to decompose organic matter into gases, predominantly CH_4 and CO_2, in anaerobic digestion processes. These processes can be augmented by dark fermentation or photofermentation, where specific bacteria directly convert organic compounds into hydrogen. Dark fermentation occurs without light, while photofermentation employs light to facilitate bacterial hydrogen production from organic substrates. Such biochemical approaches offer environmentally friendly alternatives, leveraging natural processes to produce hydrogen.

Physicochemical conversion methods, including liquid biomass steam reforming, involve reacting liquid organic compounds derived from biomass with high-temperature steam to produce hydrogen-rich syngas. This gas can then undergo purification to obtain pure hydrogen. Each of these conversion routes presents its own set of advantages, drawbacks, and technological complexities, necessitating careful consideration based on factors like biomass availability, economic viability, and sustainability goals. Overall, the overarching objective remains consistent: to optimize hydrogen production while mitigating environmental impacts and advancing toward a more sustainable energy future.

2.2.5.2　Available Technologies for Hydrogen Production from Biomass

Hydrogen production from biomass represents a significant avenue in the quest for sustainable energy solutions, leveraging various sophisticated technological approaches. Gasification, a thermochemical process, involves subjecting solid biomass to high temperatures (typically between 800°C and 1000°C) in the presence of a controlled amount of oxygen or steam. This thermochemical reaction yields syngas, a mixture primarily composed of CO and hydrogen (H_2). Syngas purification and subsequent conversion steps, such as steam reforming and WGS reactions, facilitate the extraction of high-purity hydrogen from the synthesized gas mixture.

Pyrolysis, another thermochemical method, operates under oxygen-deprived conditions, initiating the thermal breakdown of biomass constituents. This process generates a range of valuable products, including biochar, bio-oil, and gases, with hydrogen being a notable component. The bio-oil fraction obtained from pyrolysis can undergo further processing to optimize hydrogen yield, enhancing the overall efficiency of the conversion process.

Anaerobic digestion, a biochemical pathway, harnesses microbial activity to decompose biomass, particularly organic waste materials, in an oxygen-free environment. This process yields biogas, primarily comprised of CH_4 and CO_2. CH_4, a precursor to hydrogen production, can be subjected to steam reforming, enabling the separation of hydrogen gas for subsequent utilization. Additionally, biological fermentation processes involving specific bacterial strains enable the direct conversion of organic substrates into hydrogen gas through controlled biochemical reactions, albeit typically at lower conversion efficiencies compared to thermochemical pathways.

Each of these technologies offers distinct advantages and challenges, necessitating comprehensive consideration of factors such as biomass composition, process efficiency, economic viability, and environmental sustainability. By tailoring these methodologies to suit specific biomass feedstocks and operational requirements, researchers and engineers can advance the development of efficient and environmentally sound pathways for biomass-derived hydrogen production, contributing to the transition toward a low-carbon energy future.

2.2.5.3 Biomass Pretreatment Methods to Optimize its Conversion to Hydrogen

Biomass pretreatment plays a pivotal role in optimizing the conversion of biomass into hydrogen, a critical step in the quest for clean and sustainable energy sources. This pretreatment process is essential because biomass, comprising complex organic materials like cellulose, hemicellulose, and lignin, often possesses intricate structures that hinder efficient conversion into hydrogen-rich gases. Pretreatment methods aim to overcome these structural barriers by breaking down biomass components into simpler forms, thus enhancing their accessibility and reactivity during subsequent conversion processes. Mechanical pretreatment methods involve physical actions such as grinding, chopping, and milling, which reduce the size of biomass particles. This size reduction increases the surface area available for reaction, facilitating better interaction with conversion catalysts and promoting more thorough conversion of biomass into hydrogen-bearing gases. Additionally, mechanical pretreatment can improve the homogeneity of biomass feedstocks, leading to more consistent conversion outcomes.

Thermal pretreatment techniques, including torrefaction and pyrolysis, harness heat to alter the chemical composition and physical properties of biomass. Torrefaction involves heating biomass at moderate temperatures (around 200–300°C) in the absence of oxygen, reducing its moisture content and enhancing its energy density by driving off volatile compounds. Pyrolysis, on the other hand, subjects biomass to higher temperatures (up to 500°C or more) in an oxygen-free environment, leading to the decomposition of biomass components into char, oils, and gases. These pyrolysis products can be further processed to yield hydrogen-rich syngas, contributing to overall hydrogen production efficiency. Chemical pretreatment methods, such as acid hydrolysis and alkalization, employ chemical agents to modify the composition and structure of biomass. Acid hydrolysis involves treating biomass with strong acids to break down complex polysaccharides into fermentable sugars, which can then be converted into hydrogen through fermentation or gasification processes. Alkalization, on the other hand, utilizes strong bases to disrupt lignin bonds and modify cellulose structure, thereby enhancing biomass digestibility and improving hydrogen production yields.

Biological pretreatment strategies leverage the action of microorganisms or enzymes to degrade biomass components into fermentable sugars. While biological pretreatment methods are generally slower compared to chemical or thermal approaches, they offer the advantage of being more environmentally friendly and can be integrated into biorefinery processes for sustainable hydrogen production from biomass. Steam explosion pretreatment involves subjecting biomass to high-pressure and temperature steam, followed by rapid decompression, resulting in the physical rupture of biomass structures and increased accessibility of biomass components to subsequent conversion processes. This mechanical and thermal shock treatment enhances the efficiency of biomass conversion into hydrogen-rich gases. Ultimately, the choice of pretreatment method depends on factors such as the type and characteristics of biomass feedstocks, desired hydrogen production yields, process economics, and environmental considerations. By optimizing biomass pretreatment strategies, researchers and engineers can unlock the full potential of biomass as a renewable feedstock for hydrogen production, advancing the transition toward a cleaner and more sustainable energy future.

2.2.5.4 Efficiency and Technical Challenges

The production of hydrogen from biomass confronts numerous intricate technical challenges necessitating rigorous scientific exploration for their resolution. Primarily, optimizing conversion efficiency remains a fundamental hurdle. Unlike fossil fuels, biomass constituents exhibit complex structures and compositions, demanding meticulous process refinement to achieve maximal hydrogen yields. Advancing gasification, pyrolysis, and fermentation methodologies to enhance hydrogen output while mitigating energy losses presents a formidable technical endeavor requiring interdisciplinary collaboration between chemistry, engineering, and materials science disciplines.

Furthermore, the intricacies of biomass treatment underscore the multifaceted nature of this endeavor. Biomass, inherently heterogeneous, encompasses diverse organic materials with varying reactivities and physical properties. Standardizing conversion processes proves arduous due to this variability, necessitating bespoke pretreatment strategies tailored to specific biomass compositions. Consequently, scientific endeavors investigate the underlying mechanisms governing biomass deconstruction and explore innovative pretreatment techniques, such as mechanical milling, thermal torrefaction, and chemical hydrolysis, to enhance biomass digestibility and streamline hydrogen production pathways. Additionally, addressing challenges related to tar formation, contaminant removal, and system integration requires comprehensive scientific inquiry to devise robust technological solutions capable of ensuring the economic viability and environmental sustainability of biomass-derived hydrogen production on a large scale.

The energy efficiency of biomass conversion to hydrogen can vary considerably depending on the method used, the type of biomass, and the pretreatment and conversion technologies involved. Table 2.8 summarizes the estimated energy efficiency of different biomass-to-hydrogen conversion methods.

The efficiency of biomass-to-hydrogen conversion processes is intricately tied to the quality and characteristics of the biomass feedstock employed. The diverse composition and properties of biomass materials, ranging from agricultural residues to woody biomass and organic waste, exert varying impacts on hydrogen yield and conversion efficiency. Factors such as moisture content, lignin content, cellulose accessibility, and ash content can significantly influence the conversion process.

For instance, lignocellulosic biomass, which includes materials like wood chips and agricultural residues, presents both opportunities and challenges for hydrogen production. While lignocellulosic biomass offers abundant and renewable feedstock potential, its complex structure and high lignin content often require intensive pretreatment processes to facilitate effective conversion. On the other hand, organic waste streams such as food waste and municipal solid waste contain readily

TABLE 2.8
Energy Efficiency of Different Biomass-to-Hydrogen Conversion Methods

Conversion Method	Estimated Energy Efficiency (%)	Remarks
Gasification	60–70	Dependent on technology and gas cleaning
Pyrolysis	40–50	Includes losses during bio-oil conversion
Anaerobic Digestion	Low (variable)	Higher efficiency if combined methane and hydrogen production is considered
Biological Fermentation	<10	Utilizes lower temperatures and lower pressures

TABLE 2.9
Influence of Biomass Type and Quality on Conversion Yield

Factors	Description	Impact on Hydrogen Yield
Chemical Composition	Cellulose and hemicellulose enhance yield; lignin reduces it.	Variable depending on composition
Moisture Content	High water content requires more energy for vaporization, reducing efficiency.	Decreases with increasing moisture
Impurities	Heavy metals, sulfur, and chlorine may require additional purification steps.	Generally reduced by impurities
Biomass Size and Shape	Fine particle size increases the reaction surface, enhancing conversion.	Increases with better preparation
Biomass Type	Agricultural residues (rich in cellulose) are favorable; woody biomass (rich in lignin) is less favorable.	Variable depending on type
Pretreatment	Thermal, chemical, or mechanical methods can render biomass more reactive.	Generally improved with good pretreatment

fermentable organic matter, offering opportunities for anaerobic digestion processes to produce hydrogen-rich biogas. Moreover, the geographical availability and regional variations in biomass resources further impact the selection and optimization of conversion technologies. Factors such as biomass availability, transportation logistics, and regional infrastructure influence the feasibility and economic viability of biomass-to-hydrogen conversion projects. Thus, a comprehensive understanding of biomass characteristics and regional resource availability is essential for designing efficient and sustainable biomass conversion processes for hydrogen production. Table 2.9 summarizes the key factors influencing hydrogen yield during the conversion of different types of biomass, highlighting the specificities and challenges associated with each factor.

2.2.5.5 Economic and Environmental Impacts

The production of hydrogen from biomass is often regarded as a promising avenue for reducing carbon emissions compared to fossil fuel-based methods. However, it comes with its own set of environmental impacts that must be carefully managed. One significant consideration is the emission of GHGs. While biomass-derived hydrogen has the potential to substantially decrease GHG emissions, achieving carbon neutrality, emissions may occur throughout the biomass supply chain, including cultivation, harvesting, transportation, and processing. Conversion processes such as gasification and pyrolysis, though contributing to CO_2 emissions, typically emit fewer pollutants than fossil fuels. This underscores the environmental advantage of biomass as a renewable energy source, despite the challenges in managing emissions.

Water consumption is another critical concern associated with biomass production for energy. Cultivating biomass crops often requires significant irrigation, which can strain local water resources, particularly in regions where water is scarce. This heightened demand for water may exacerbate existing water management issues and lead to conflicts between agricultural, industrial, and domestic water needs. It emphasizes the importance of implementing sustainable agricultural practices and optimizing water usage in biomass production to mitigate environmental impacts and ensure the long-term sustainability of these energy endeavors. Land use implications pose substantial challenges for biomass production, particularly regarding competition with food agriculture and environmental degradation. The conversion of land for biomass cultivation may compete with food production, potentially affecting food security and socioeconomic dynamics. Furthermore, unsustainable land management practices risk deforestation and biodiversity loss, compromising local ecosystems' integrity and the services they provide. To address these concerns, responsible and sustainable approaches are necessary, balancing energy production with natural resource conservation and food security objectives. To make hydrogen production from biomass more efficient and environmentally friendly, numerous developments and innovations are underway across various domains. Here are some key areas of advancement.

2.2.5.5.1 *Improvement of Conversion Technologies*

Enhancing biomass-to-hydrogen conversion technologies involves several key developments. Advanced gasification is undergoing extensive research to create more efficient systems that reduce tar formation while increasing hydrogen yield. Catalytic pyrolysis utilizes catalysts to enhance biomass decomposition and increase hydrogen production. Additionally, fermentation is being enhanced through the development of microbial strains, including genetically modified ones, that are more efficient in hydrogen production. These advancements aim to optimize the conversion of biomass into a clean and efficient energy source.

2.2.5.5.2 *Carbon Capture and Utilization (CCU)*

Carbon capture and utilization (CCU) in hydrogen production encompass two main aspects. First, CO_2 sequestration involves integrating advanced CO_2 capture technologies to make hydrogen production practically carbon-neutral. Second, the utilization of captured CO_2 involves developing methods to convert CO_2 into useful chemicals or fuels. This approach not only increases the overall profitability of the process but also helps reduce the environmental impact of industrial production.

2.2.5.5.3 Optimization of Biomass Pretreatment

Optimizing biomass pretreatment focuses on developing innovative techniques that are both less energy-intensive and more effective. These methods aim to prepare biomass optimally before its conversion into useful products. Among these techniques are enzymatic treatments that target biomass decomposition and processes using advanced solvents that facilitate the separation of desirable components. These improvements maximize conversion efficiency while minimizing energy consumption and associated costs.

2.2.5.5.4 Hybridization of Technologies

Hybridization of technologies for hydrogen production relies on two main approaches to increase efficiency and decrease environmental impacts. The first approach involves combined systems that integrate different hydrogen production technologies, such as gasification and fermentation. This combination aims to leverage the specific strengths of each method to improve the overall efficiency of the process. The second approach involves integration with renewable energies, using renewable energy sources to power biomass conversion processes, thereby helping reduce the carbon footprint of the system. These strategies are essential to make hydrogen production more sustainable and environmentally friendly.

2.2.5.5.5 Management and Valorization of By-products

Managing and valorizing by-products in biomass-based hydrogen production encompass two crucial aspects. The first concerns the use of biochar, where research is being conducted to explore new potential applications, including advanced construction materials or energy storage devices. This work aims to exploit the unique properties of biochar for innovative uses that can contribute to environmental sustainability. The second aspect is the valorization of process residues, which involves transforming these residues into value-added products such as bioplastics or composite materials. This approach not only reduces waste but also promotes the creation of useful products, adding an economic dimension to by-product management.

2.2.5.5.6 Sustainable Supply Chain Models

Sustainable supply chain models for biomass-based hydrogen production include specific strategies to optimize logistics and encourage responsible agriculture. On one hand, biomass logistics focuses on optimizing biomass collection, transportation, and storage, with goals to reduce costs and minimize environmental impact. This optimization ensures maximum efficiency while preserving natural resources. On the other hand, the development of sustainable agricultural practices aims to maximize biomass production while reducing negative environmental impacts such as deforestation and overexploitation of water resources. These practices include adopting agricultural techniques that support soil health, water conservation, and biodiversity, thus contributing to a sustainable biomass supply for hydrogen production.

These innovations aim not only to improve the efficiency and profitability of biomass-based hydrogen production but also to mitigate associated environmental impacts, supporting a more sustainable energy transition. Accordingly, hydrogen production from biomass is a method that utilizes renewable resources to produce hydrogen through processes such as gasification, pyrolysis, or fermentation. This production is often more expensive than traditional methods using fossil fuels, but it offers environmental benefits that could justify the additional costs in the context of sustainable development policies and GHG emission reduction. Table 2.10 is a comparative table of hydrogen production costs by different methods, including production from biomass.

2.2.5.6 Policies and Market Viability

Energy policies play a pivotal role in either fostering or impeding the advancement of hydrogen production from biomass. Governments have various means to bolster this sector, ranging from direct subsidies and low-interest loans to tax incentives and mandates for renewable hydrogen adoption in key industries. Additionally, funding research endeavors into biomass gasification and hydrogen purification processes aids in surmounting technical hurdles and driving down costs. By establishing regulatory frameworks specific to hydrogen and promoting renewable energy sources more broadly, policymakers can create an environment conducive to the growth of biomass hydrogen production. Conversely, barriers to the expansion of hydrogen from biomass are multifaceted and substantial. The absence of clear regulatory guidelines tailored to hydrogen may deter investors and hinder technological adoption. Competition from other energy sources, such as fossil fuels or established renewables like wind and solar, may divert resources and attention away from biomass hydrogen development. Moreover, stringent environmental regulations and sustainability considerations, including land use constraints for biomass cultivation, could pose challenges to the viability of this hydrogen production method. Complex

---

TABLE 2.10

Comparison of Hydrogen Production Costs from Biomass Using Different Methods

Production Method	Approximate Cost per Kilogram of Hydrogen (USD)	Comments
Hydrogen from Biomass	2.5–6	Depends on the type of biomass and conversion technology
Water Electrolysis (Renewable Energy)	3–7	High costs due to the price of renewable electricity
Water Electrolysis (Conventional Energy)	1–4	Less expensive than conventional electricity, but environmental impact
Natural Gas Reforming	1–2	The most economical method currently, but depends on the price of natural gas
Coal Gasification	1.5–3	Effective but high CO_2 emissions and depends on the price of coal

permit processes for facility construction and operation further complicate matters, potentially causing delays or dissuading project involvement due to lengthy approval procedures.

Hydrogen derived from biomass offers a promising avenue for reducing reliance on fossil fuels, presenting several key advantages in its production and application. First, its renewability sets it apart from finite fossil fuel sources, as biomass originates from organic materials like agricultural and forestry waste, providing a sustainable alternative. Second, biomass-derived hydrogen production can be environmentally advantageous by potentially achieving carbon neutrality, particularly if CO_2 emissions during production are captured and stored or repurposed. This contrasts starkly with fossil fuel extraction and combustion, which release previously sequestered carbon into the atmosphere. Additionally, biomass hydrogen can seamlessly integrate with existing technologies such as fuel cells and industrial processes, facilitating a smoother transition away from fossil fuels.

However, several challenges must be addressed to fully realize the potential of biomass-derived hydrogen. First, improvements in efficiency and cost competitiveness are necessary for biomass hydrogen production technologies to rival fossil fuels and other renewable energy sources economically. Second, the availability of sustainable biomass is finite, and overuse could lead to environmental degradation or competition with essential land uses like food production. Lastly, significant investments are required to develop distribution infrastructures that can efficiently deliver hydrogen to end users, ensuring its widespread adoption and utilization. Accordingly, while biomass-derived hydrogen holds immense promise for reducing fossil fuel dependency, its realization hinges on overcoming economic, technical, and logistical hurdles. Effective energy policies and strategic investments are crucial for navigating these challenges and fostering the transition toward a more sustainable energy future.

The stability and availability of biomass supply are crucial for the viability of hydrogen production from this resource. These factors directly influence the feasibility and efficiency of this type of hydrogen production. First, supply stability entails a consistent and predictable source of raw materials, necessary for continuous operations. Interruptions in biomass supply can lead to production shutdowns, increasing operational costs and reducing overall efficiency. Second, a stable source allows for better operational planning, optimized biomass stock management, and reduced costs related to storage and logistics. Availability of biomass refers to the quantity that can be sustainably harvested or generated without compromising environmental, social, or economic needs. Key aspects include volume, renewal, seasonality, regional variation, and competition. Biomass must be available in sufficient quantities to meet production demand, with renewable growth essential for long-term sustainability. Seasonal fluctuations and regional variations in biomass type and quantity can influence the location and capacity of hydrogen production facilities. Additionally, competition for biomass across various sectors, including energy production and biochemical manufacturing, may limit availability and influence prices. Thus, ensuring a stable and abundant biomass supply is crucial for maintaining predictable and competitive raw material costs and justifying infrastructure investments for biomass collection, transportation, and conversion into hydrogen. In conclusion, the viability of hydrogen production from biomass heavily relies on securing a stable and adequate biomass supply.

Fluctuations in availability can pose significant operational and economic challenges, necessitating well-designed planning and policies to ensure the sustainability and efficiency of this renewable energy source.

The deployment of large-scale hydrogen production from biomass faces several regulatory obstacles that need to be addressed to facilitate its development. The main regulatory challenges include stringent environmental standards, such as GHG emission regulations and waste management rules, which may pose challenges for biomass-based hydrogen production facilities, particularly if emission control technologies are not fully developed or integrated. Additionally, regulations governing sustainable biomass harvest and land use competition for other purposes, like biofuel production, require a delicate balance to ensure environmental sustainability and food security without compromising land prices. Certification and quality standards for hydrogen production, alongside safety regulations for facilities, also present regulatory hurdles. Hydrogen produced must meet specific specifications for commercial applications, necessitating clearly defined quality standards and certifications, which can be a lengthy and costly process. Furthermore, integrating production facilities into existing energy infrastructures poses challenges related to network compliance and regulatory requirements, including interconnection regulation and right-of-way issues. Governmental support, through subsidies and financial incentives, tailored to biomass hydrogen may be necessary to foster innovation and overcome regulatory barriers. Overall, adapting and improving the current regulatory framework is essential for the viability of large-scale hydrogen production from biomass, requiring flexibility in restrictions, specific standards for biomass hydrogen, and incentives to ensure environmental protection and public safety.

2.2.5.7 Future Perspectives and Stakeholder Engagement

Hydrogen derived from biomass holds tremendous potential in shaping the future energy landscape, offering a range of benefits across environmental, economic, and social dimensions. Its capacity to reduce GHG emissions, particularly when coupled with CCS technologies, positions it as a key player in mitigating climate change. Additionally, biomass-derived hydrogen offers a pathway to diversify energy sources, reducing reliance on finite fossil fuels and enhancing energy security. Its versatility extends to applications across electricity generation, heating, transportation, and industrial processes, contributing to a more sustainable and resilient energy system. Moreover, the production of hydrogen from biomass can stimulate local economies, especially in rural areas rich in biomass resources. By fostering job creation in biomass collection, processing, and hydrogen production, it offers opportunities for economic growth and community development. Furthermore, the integration of biomass-derived hydrogen into various sectors, such as agriculture, industry, and transportation, fosters innovation and promotes the transition toward cleaner technologies. However, realizing the full potential of biomass-derived hydrogen hinges on continued investment in research, development, and infrastructure, alongside the implementation of supportive policies and regulatory frameworks to facilitate its adoption and scale-up.

Research and development of hydrogen derived from biomass involve various key stakeholders, including academic institutions, businesses, research organizations, and governments worldwide. Leading countries in this field include the United States, which has made significant investments in renewable energy research, particularly focused on biomass-derived hydrogen. Institutions such as the National Renewable Energy Laboratory (NREL) and various universities are actively engaged in research aimed at optimizing biomass conversion processes into hydrogen. Germany, with its strong government policies and research investments, is also a leader in hydrogen and renewable energy technologies. Projects supported by institutions like the Fraunhofer Institute are dedicated to improving methods for producing green hydrogen. Japan is another significant player, boasting a well-established hydrogen strategy and investments in various hydrogen technologies, including those based on biomass. The country views hydrogen as a crucial element of its future energy strategy, especially in reducing its external energy dependence. Canada, rich in forest and agricultural resources, actively explores hydrogen production from biomass sources. Research projects conducted by institutions like the University of British Columbia, in collaboration with industrial partners, contribute to advancing biomass-derived hydrogen technology. Additionally, China is investing in biomass-derived hydrogen technology as part of its carbon neutrality ambitions, with a particular focus on biomass gasification in rural areas.

Key actors in this field include technological and energy companies like Siemens, Bosch, and specialized startups, which are developing technologies related to biomass-derived hydrogen production, storage, and utilization. Academic and research institutions worldwide, particularly in Europe, the United States, and Asia, play a crucial role in conducting research on fundamental and applied aspects of hydrogen production from biomass. International organizations such as the IEA and the EU also contribute to research and international collaboration through programs like Horizon Europe. Continuous engagement in research and development is essential for overcoming technical and economic challenges and integrating biomass-derived hydrogen technology into the global energy mix sustainably and effectively.

Also, to promote innovation and adoption of biomass-derived hydrogen production technology, forging strategic partnerships among various stakeholders in the sector is essential. These collaborations can facilitate knowledge sharing, optimize resources, and expedite the development and commercialization of new technologies. Industry–university partnerships enable combining academic expertize with industrial resources for advanced research and training programs, ensuring a skilled workforce for the industry. Public–private partnerships allow governments to finance research and development of sustainable hydrogen technologies, while collaborations between biomass hydrogen producers and transportation industries develop hydrogen mobility solutions. International cooperation facilitates technology exchanges and joint research programs to address global challenges. Alliances with NGOs provide regulatory support and promote public awareness, while industrial consortia allow risk and resource sharing for large-scale development. Such strategic partnerships are crucial for overcoming technological, economic, and regulatory obstacles, accelerating the adoption of biomass-derived hydrogen as part of the global energy transition.

Improving public awareness and acceptance of biomass-derived hydrogen production is essential for supporting the development and deployment of this technology. Several effective strategies include educational programs in schools, universities, and communities to teach the basics and benefits of biomass-derived hydrogen, along with media and online campaigns disseminating clear information about its environmental and economic advantages. Community engagement through public consultations and involvement in project planning fosters understanding and support. Transparency is key, with regular updates on project development and risk communication to build public trust. Demonstrations and pilot projects provide tangible examples of the technology's benefits, while collaboration with opinion leaders and influencers helps broaden outreach. Highlighting economic benefits such as job creation and offering financial incentives further incentivizes adoption. Political support and a favorable regulatory framework bolster public confidence. In conclusion, a multidimensional approach combining education, engagement, transparency, technology demonstration, and policy support is crucial for biomass-derived hydrogen to become a cornerstone of a sustainable energy future.

2.2.6 Pyrolysis

Pyrolysis, used for hydrogen production from hydrocarbons, petrochemical residues, biomass, and organic waste, represents a sophisticated thermochemical approach that converts these raw materials into hydrogen and other valuable products. This technique involves heating these materials to high temperatures in an oxygen-free environment, resulting in their chemical decomposition without combustion. The process releases a gas mixture, including hydrogen, along with solid and liquid residues that can also be utilized. Strategically employed, pyrolysis can significantly contribute to the energy transition by offering a method for waste valorization and reducing dependence on fossil fuels. Moreover, this method can be adapted to operate at different scales, making it applicable to both industrial installations and smaller-scale projects, closer to biomass or waste sources.

The world faces urgent pressure to reduce GHG emissions and limit climate change. In this context, hydrogen is seen as a key energy carrier, capable of storing and transporting energy without direct CO_2 emissions during its use. However, most of the hydrogen produced today comes from fossil sources, particularly natural gas, through processes like steam reforming, which are themselves significant sources of CO_2 emissions. Pyrolysis offers a promising alternative as it can convert nonfossil raw materials, such as biomass and organic waste, into hydrogen, with a potentially neutral or even negative carbon footprint if solid by-products like biochar are used for carbon sequestration. On the technological front, pyrolysis for hydrogen production is at the convergence of several major advances in chemical engineering, materials, and energy management. Challenges include improving reactor efficiency, managing extreme temperatures, and developing catalysts that can increase hydrogen yield while minimizing the formation of undesirable by-products. The significant-scale implementation of pyrolysis also depends on economic and social factors. The initial costs of pyrolysis technologies are high, and their profitability can be challenging to achieve without political support and subsidies. Additionally, the localization of pyrolysis facilities must be strategically planned to be close to biomass sources and waste treatment centers, posing logistical and community acceptance issues. Environmental policies play a crucial role in the deployment of technologies like pyrolysis. Regulations that encourage waste reduction, recycling, and the use of clean technologies can accelerate the adoption of pyrolysis. At the same time, international emission standards and subsidies for green technologies directly influence investments in this sector. In summary, pyrolysis for hydrogen production fits into a multidimensional context where environmental, technological, economic, social, and regulatory considerations intersect. Its future development will depend not only on technical progress but also on the evolution of policy frameworks and market dynamics.

2.2.6.1 Technical and Scientific Challenges

The technical and scientific challenges associated with pyrolysis for hydrogen production are numerous and complex, reflecting the inherent difficulties in implementing advanced technology within an industrial and environmental framework. Here's a detailed exploration of these challenges:

- Process Complexity: Pyrolysis demands substantial energy input to reach temperatures ranging from 400°C to 1400°C, essential for stability given diverse feedstock types with distinct chemical properties. Moreover, the introduction of catalysts to enhance hydrogen yield poses challenges, including high operational costs and the need for frequent regeneration or replacement due to catalyst degradation or "poisoning" by impurities in the feedstock. While solid residues like biochar and coke from pyrolysis can have useful applications, their management must be ecologically and economically viable. This entails considering logistics, target markets, and environmental regulations to prevent issues like soil and water contamination, emphasizing environmentally friendly management practices.
- Optimization of Operational Conditions: Tuning parameters such as temperature, pressure, and residence time in the pyrolysis reactor is essential for maximizing hydrogen production efficiency. Advanced control systems allowing real-time adjustments and continuous monitoring are crucial, along with ongoing research and development efforts to understand reaction mechanisms and develop more efficient catalytic materials. Overcoming these challenges is paramount to making pyrolysis a viable method of hydrogen production on a large scale.

In hydrogen production by pyrolysis, the most commonly used method is fast pyrolysis, converting biomass solids into liquid bio-oil with high yield, solid biochars, and noncondensable gases. Another method, plasma pyrolysis, is notable for its potential in carbon-free hydrogen production from CH_4, although not as widely used as fast pyrolysis in current industrial applications. These methods offer significant benefits in waste reduction, resource management, and renewable energy production, aligning with decarbonization and sustainable resource utilization goals.

2.2.6.2 Applied Materials

For hydrogen production through pyrolysis, three types of materials are commonly used: biomass, hydrocarbons, and plastic waste. A comparative and quantitative summary is presented in Table 2.11. This comparison illustrates that while all three

TABLE 2.11

Comparison of hydrogen production from different materials, using pyrolysis method

Material	Conversion Efficiency to Hydrogen	Pyrolysis Temperature	Hydrogen Yield	Typical Applications
Biomass	Average to High	400°C–1000°C	5%–20% of biomass weight	Renewable energy production, biofuels
Hydrocarbons	High	650°C–1000°C	Up to 30% of hydrocarbon weight	Chemical industry, petroleum refining
Plastic Waste	Variable depending on plastic type	300°C–900°C	Up to 64% more effective than tires	Plastic recycling, fuel production

materials are viable for hydrogen production through pyrolysis, the choices often depend on desired efficiency, material availability, and specific hydrogen production goals. Plastics and hydrocarbons generally offer higher hydrogen yields than biomass, but their availability and environmental impact can vary significantly.

It is noteworthy to mention that the key steps in the pyrolysis process for hydrogen production involve rapid heating of organic materials to temperatures ranging from 400°C to 1400°C, crucial for complete breakdown and maximum hydrogen release. This is followed by thermal decomposition, where complex chemical bonds break down, yielding a gas mixture containing hydrogen, along with other gases, liquids, and solid residues. Gas purification is then carried out to separate hydrogen from other gases, resulting in high-purity hydrogen suitable for various energy applications. Specific hydrogen yields vary, with biomass pyrolysis typically yielding 10 to 20% of its weight in hydrogen, depending on composition and process conditions, while optimized systems can achieve conversion efficiencies of up to 60–70%. Hydrogen production by pyrolysis presents an alternative to conventional methods like steam reforming of hydrocarbons and water electrolysis, offering potential sustainability benefits if feedstocks are renewable or waste. Advantages include CO_2 emissions reduction, waste valorization, and decentralized production, but challenges include high investment costs and variable yields. Quantitative information shows significant variability in hydrogen yields, highlighting the need for optimized processes. Overall, while promising, pyrolysis for hydrogen production requires addressing technical, economic, and regulatory hurdles to realize its full potential.

2.3　EMERGING LESS COMMON METHODS

2.3.1　Photoelectrolysis of Water

Water photoelectrolysis is a promising method for hydrogen production, leveraging solar energy to split water molecules into hydrogen and oxygen. It operates through a series of steps: first, semiconductors acting as photoelectrodes absorb sunlight, exciting electrons and creating electron-hole pairs. These separated charges then facilitate the electrolysis reaction, where electrons reduce water to form hydrogen gas while holes generate oxygen gas. The gases are collected separately, with potential for further applications like fuel cells or energy storage. Despite its environmental benefits, such as zero CO_2 emissions and reliance on abundant resources like water and sunlight, challenges remain in optimizing photoelectrode efficiency and reducing manufacturing costs, hindering widespread adoption. Efforts in advancing water photoelectrolysis technology are crucial for sustainable hydrogen production, as it offers a clean and renewable energy pathway. Improvements in materials science and engineering are needed to enhance the efficiency and scalability of this process, making it economically viable for large-scale applications. Overcoming these challenges could unlock the full potential of water photoelectrolysis as a key contributor to a low-carbon energy future, supporting endeavors to combat climate change and transition toward cleaner energy sources.

2.3.2 Biological Hydrogen Production (Biohydrogen)

Biological hydrogen production, also known as biohydrogen, relies on microorganisms like bacteria or algae to generate hydrogen via fermentation or photosynthesis. Anaerobic fermentation involves certain bacteria breaking down organic compounds in the absence of oxygen, producing hydrogen as a metabolic by-product. These bacteria can utilize diverse organic substrates such as sugars, agricultural and industrial wastes, or wastewater. Similarly, certain microalgae and cyanobacteria possess the ability to produce hydrogen through biological photosynthesis, converting CO_2 and water into carbohydrates and oxygen, with hydrogen produced as a by-product, particularly under stress conditions like high light levels or low nutrient availability. Additionally, biohydrogen can be derived from organic waste materials such as agricultural residues, crop remnants, or sewage sludge through anaerobic fermentation processes. Microbial consortia present in these wastes biodegrade the organic matter, yielding hydrogen as a by-product. Research into genetic engineering of microorganisms aims to further enhance biological hydrogen production efficiency by modifying their genetic makeup to optimize hydrogen production pathways. Despite its potential benefits, including the utilization of renewable resources and the absence of CO_2 emissions, challenges persist in improving production efficiency and managing undesirable by-products.

2.3.3 Natural Hydrogen Production in Underground Reservoirs

The generation of hydrogen in underground reservoirs, often referred to as "natural hydrogen production" or "geological hydrogen," is a naturally occurring process whereby hydrogen is produced and stored within geological formations. This phenomenon arises from various sources, including chemical reactions in the Earth's crust and biological activities. One significant chemical reaction is serpentinization, where ultramafic rocks rich in olivine interact with water, resulting in the formation of molecular hydrogen (H_2) and serpentine minerals. Additionally, the radiolysis of water, driven by natural radiation (alpha, beta, and gamma) from the decay of radionuclides in rocks, can also lead to hydrogen production. Furthermore, certain microorganisms, known as hydrogenogens, can generate hydrogen through biochemical processes like fermentation and methanogenesis.

Hydrogen produced deep within the Earth's crust can migrate through rock fractures and faults. This movement is facilitated by the low density and small molecular size of hydrogen, enabling it to traverse rock pores more easily. Once generated, hydrogen can accumulate in underground reservoirs akin to those of hydrocarbons, such as oil and natural gas. Typically, these reservoirs consist of porous rock formations, like sandstones, overlaid by impermeable rock layers, such as clays, which serve as natural traps. Hydrogen reservoirs can be classified into two types: primary and secondary. Primary reservoirs are those where hydrogen is produced and accumulated in situ through geochemical or biological reactions. Conversely, secondary reservoirs are characterized by hydrogen that has migrated from other deeper formations and accumulated in suitable geological structures.

The exploration for natural hydrogen employs geophysical and geochemical techniques similar to those used in hydrocarbon exploration, including seismic, magnetotelluric, and soil gas surveys. The extraction of hydrogen from these reservoirs can be achieved through drilling methods similar to those used in the oil industry. However, the infrastructure and techniques for storing and transporting hydrogen must be tailored to its unique properties, such as low density and high reactivity.

Several natural hydrogen reservoirs have been discovered worldwide. In the Bourakébougou region of Mali, for instance, a well drilled by the Petroma company uncovered hydrogen at high concentrations (up to 98%). In Russia, the Kola region is noted for its natural hydrogen manifestations, with emanations detected in springs and seeps. In the United States, traces of natural hydrogen have been identified in various regions, including Kansas. Canada's Canadian Shield region has shown evidence of natural hydrogen, particularly in the fracture zones of Precambrian rocks. In South Africa, research has indicated the presence of natural hydrogen within fault systems and fractures. In Australia, studies have highlighted the potential for natural hydrogen in certain geological formations. Similarly, in Brazil, natural hydrogen manifestations have been observed in the São Francisco region.

The production of hydrogen in underground reservoirs involves a complex interplay of chemical and biological reactions within the Earth's crust. Understanding and harnessing these resources could offer a significant source of hydrogen, supporting the shift toward cleaner and more sustainable energy sources. Accordingly, a roadmap illustrating the challenges and solutions for implementing underground hydrogen storage in hydrogen production is shown in Figure 2.7.

2.3.4 Methanolysis

The methanolysis method for hydrogen production involves converting CH_4 into hydrogen and CO_2 using water or steam as a methanolysis agent. This chemical process occurs through the reaction: $CH_4 + H_2O$ (or vaporized H_2O) $\rightarrow CO_2 + 3H_2$. It's an exothermic reaction, releasing heat as it converts one molecule of CH_4 and one molecule of water into one molecule of CO_2 and three molecules of hydrogen. To optimize this process, methanolysis is typically conducted at high temperatures

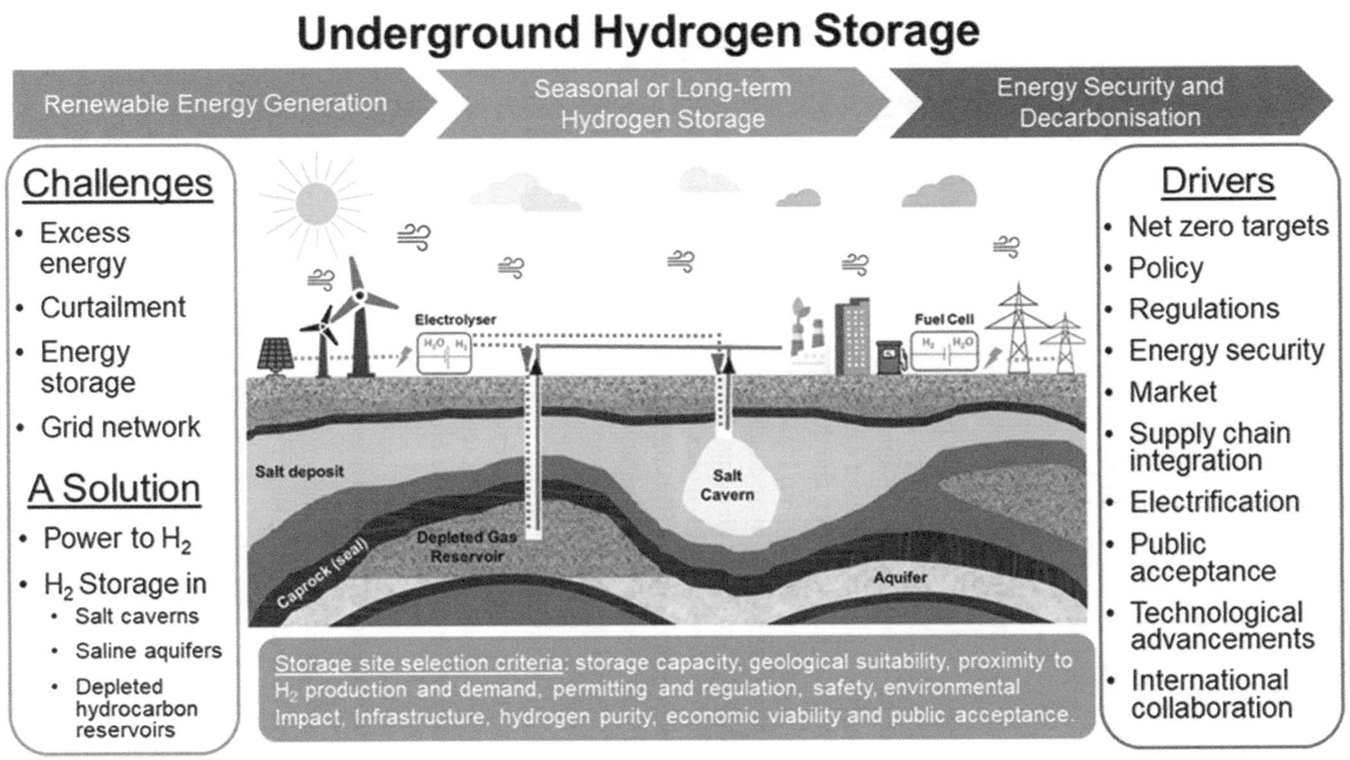

FIGURE 2.7 Methods of hydrogen production.

(Adapted from ref. [8].)

ranging between 700°C and 1000°C, often under pressure, to encourage the dissociation of CH_4 and water into their constituent elements. Catalysts like nickel, cobalt, ruthenium, or metal alloys are commonly employed to facilitate CH_4 and water dissociation, promoting efficient hydrogen formation. This method finds application in harnessing hydrogen from CH_4-rich sources such as natural gas, biogas, or organic waste, offering an alternative to CH_4 reforming and a means of utilizing residual gases from industrial processes. By using CH_4 as a hydrogen source and producing CO_2, methanolysis enables the capture and storage of CO_2 to mitigate GHG emissions. Despite its benefits, challenges persist, including the need for high reaction conditions, potential reliance on costly catalysts, and the technical and economic complexities associated with separating and purifying the produced hydrogen.

2.3.5 AMMONIA DECOMPOSITION

The methanolysis method for hydrogen production involves chemically converting CH_4 into hydrogen (H_2) and CO_2 using water or steam as a methanolysis agent. This process, represented by the general reaction $CH_4 + H_2O$ (or vaporized H_2O) $\rightarrow CO_2 + 3H_2$, is exothermic, releasing heat as one molecule of CH_4 and one molecule of water transform into one molecule of CO_2 and three molecules of hydrogen. Typically conducted at high temperatures ranging between 700°C and 1000°C and often under pressure, methanolysis requires these conditions to encourage the dissociation of CH_4 and water into their constituent elements. Catalysts like nickel, cobalt, ruthenium, or metal alloys are frequently employed to expedite CH_4 and water dissociation, facilitating efficient hydrogen formation. This method finds application in harnessing hydrogen from CH_4-rich sources such as natural gas, biogas, or organic waste, offering an alternative to CH_4 reforming and a means of utilizing residual gases from industrial processes. By utilizing CH_4 as a hydrogen source and simultaneously producing CO_2, methanolysis allows for the capture and storage of CO_2 to mitigate GHG emissions. However, its implementation necessitates high reaction conditions and may involve the use of costly catalysts. Additionally, technical and economic challenges may arise in the separation and purification of the produced hydrogen.

2.3.6 HYDROCARBON DECOMPOSITION

The process of hydrogen production through hydrocarbon decomposition, such as ethane, typically involves either steam reforming or pyrolysis, depending on the chosen method. SMR is the predominant approach, where hydrocarbons like CH_4 or ethane are mixed with steam and heated in the presence of a catalyst, yielding hydrogen and CO. The subsequent WGS reaction further converts CO into additional hydrogen and CO_2. Following this, hydrogen is separated and purified from other gases, primarily CO_2, for utilization or storage. Alternatively, pyrolysis offers an alternative route, involving heating the hydrocarbon at high temperatures without oxygen, leading to the decomposition of ethane into hydrogen and solid carbon. Despite its potential for producing pure hydrogen and minimizing CO_2 emissions, pyrolysis poses challenges in managing solid carbon residues and its high energy demands. Both methods offer distinct advantages and challenges. Steam reforming is widely utilized due to its efficiency and established infrastructure, albeit generating CO_2 as a by-product. Pyrolysis, on the other hand, presents a cleaner approach with no CO_2 emissions, yet it necessitates intricate carbon handling and consumes significant energy. Balancing environmental concerns with practical feasibility is crucial in determining the most suitable method for hydrogen production from hydrocarbons, considering factors such as resource availability, technological advancements, and environmental impact mitigation strategies.

2.3.7 Electrochemical Methanol Decomposition

The electrochemical decomposition of methanol for hydrogen production stands as a compelling avenue for clean energy generation. Utilizing electrochemical reactions, this method breaks down methanol into hydrogen and CO_2 under controlled conditions. Methanol, chosen for its liquid state at ambient temperatures, serves as the primary reactant within an electrochemical cell equipped with an electrolyte and two electrodes—anode and cathode. In this process, methanol undergoes oxidation at the anode, generating protons, electrons, and CO_2. The released electrons then traverse through the external circuit to the cathode, where protons combine with them to form gaseous hydrogen. This multistep process represents a significant stride toward sustainable hydrogen production, offering potential solutions to environmental concerns associated with traditional methods. Despite its promise, electrochemical methanol decomposition presents certain challenges. One such challenge involves managing the CO_2 by-product, necessitating effective capture and utilization strategies to mitigate environmental impact. Additionally, the development of efficient and durable catalysts for methanol oxidation and hydrogen production reactions remains crucial for enhancing process efficiency and economic viability. Nevertheless, the method's potential applications span various sectors, including fuel cells, chemical synthesis, and renewable energy storage, signifying its role in shaping a cleaner and more sustainable energy landscape.

2.3.8 Thermochemical Decomposition of Ammonia

The thermochemical decomposition of ammonia (NH_3) represents a promising avenue for clean hydrogen production, leveraging heat to disassemble ammonia molecules into their elemental components, hydrogen (H_2), and nitrogen (N_2). This method, often heralded for its efficiency and transportability, offers a glimpse into a sustainable future, where hydrogen can be harnessed as a versatile energy carrier. At its core, the process embodies an intricate dance of chemical bonds, where ammonia molecules, subjected to elevated temperatures, gracefully yield hydrogen and nitrogen gases, akin to unraveling a complex tapestry to reveal its constituent threads.

In the heart of the decomposition lies a delicate interplay of heat and chemistry. Ammonia, warmed to temperatures ranging from 500°C to 850°C, undergoes a metamorphosis, shedding its molecular identity to birth hydrogen and nitrogen gases. Catalysts, acting as silent conductors in this symphony of reactions, orchestrate the process, reducing the energy threshold required for transformation. This method boasts a multitude of advantages, from the unparalleled purity of the resulting hydrogen, akin to the clarity of mountain spring water, to the practicality of transporting liquefied ammonia, likened to the seamless flow of a river through diverse landscapes. Yet, it faces its share of challenges, from the voracious energy appetite demanded by the endothermic nature of the reaction to the meticulous quest for catalysts resilient enough to withstand the rigors of high-temperature environments. As research strides forward, navigating these intricacies, the horizon gleams with the promise of a sustainable energy paradigm, where ammonia's metamorphosis fuels a world powered by clean hydrogen.

2.3.9 THERMOCHEMICAL DECOMPOSITION OF WATER

Thermochemical water decomposition for hydrogen production involves utilizing high temperatures to catalyze the splitting of water (H_2O) into oxygen (O_2) and hydrogen (H_2). Unlike electrolysis, which employs electricity to separate hydrogen and oxygen, thermochemical processes leverage heat, often sourced from renewable or nuclear energy, to induce a series of chemical reactions releasing hydrogen from water. This method is particularly appealing for its capacity to harness the elevated temperatures found in specific industrial or energy settings, potentially enabling more efficient and cost-effective hydrogen production. The operation of thermochemical decomposition encompasses multiple thermochemical reactions, typically a sequence of intermediate chemical reactions instead of a straightforward decomposition reaction. These reactions are carefully selected so that, at the cycle's conclusion, hydrogen and oxygen are the sole products, with all intermediate reactants regenerated. Temperatures exceeding 500°C are usually required, sometimes reaching 800°C or higher, with heat sourced from solar, geothermal, or surplus heat from a nuclear reactor. Various thermochemical cycles have been proposed and studied, like the IS cycle, the iron–chlorine cycle, and the copper–chlorine cycle, each with specific reactions and operating conditions. For instance, the IS cycle entails three main reactions and demands temperatures of up to 850°C for its most thermally intensive step.

2.3.10 HIGH-TEMPERATURE ELECTROLYSIS

HTE using SOEC is an advanced technology for hydrogen production, distinguished by its ability to operate at elevated temperatures, typically between 700°C and 800°C. This characteristic enhances the overall efficiency of the electrolysis process, thereby reducing the energy cost associated with hydrogen production. SOECs consist of three main components: an anode, a cathode, and a solid electrolyte that separates them. The electrolyte, typically made of YSZ, enables the conduction of oxygen ions at high temperatures. At the cathode, water vapor (H_2O) is reduced to hydrogen gas (H_2) and oxygen ions (O_2^-), with the necessary electrons provided by an external electric current. These oxygen ions then migrate through the solid electrolyte to the anode, where they combine with electrons from the external circuit to form gaseous oxygen, completing the electrical circuit. The use of high temperatures in SOECs reduces the overpotential required for electrolysis, enhancing energy efficiency. Additionally, their versatility allows operation in reverse mode as Solid Oxide Fuel

Cells (SOFCs) for electricity production from hydrogen, and their compatibility with industrial waste heat sources further improves overall energy efficiency. However, challenges such as material durability and high initial costs remain to be addressed for the widespread deployment of SOEC technology.

2.3.11 HIGH-PRESSURE ELECTROLYSIS

High-pressure electrolysis represents a significant advancement in hydrogen production technology, offering the capability to produce hydrogen directly at elevated pressures. Unlike conventional electrolysis conducted at atmospheric pressure, high-pressure electrolysis eliminates the need for additional mechanical compression of hydrogen, thereby reducing operational costs and energy consumption. This method operates on the same fundamental principle as standard electrolysis, where water molecules are split into oxygen and hydrogen using an electric current. However, in high-pressure electrolysis, these reactions occur within a pressurized chamber, typically ranging between 30 and 200 bars, ensuring that the hydrogen produced is already at high pressure, ready for storage or transportation without further compression. One of the key advantages of high-pressure electrolysis is its potential to significantly reduce the costs associated with hydrogen compression. By producing hydrogen at high pressure directly during the electrolysis process, the need for additional mechanical compression is minimized or eliminated, resulting in cost savings and improved energy efficiency. Additionally, the enhanced safety of handling hydrogen at high pressure, coupled with the ability to integrate high-pressure electrolysis systems with renewable energy sources, makes this technology an attractive option for advancing the hydrogen economy and supporting the transition to clean energy. However, challenges related to investment costs, thermal management, and durability of electrolyzer components remain important considerations for the widespread adoption of high-pressure electrolysis in industrial-scale hydrogen production.

2.3.12 HYDROGEN PEROXIDE DECOMPOSITION

The decomposition of hydrogen peroxide (H_2O_2) stands as a straightforward chemical pathway for hydrogen production, notable for its simplicity and rapid reaction kinetics. While H_2O_2 naturally decomposes into water and oxygen, the addition of specific catalysts alters this process to favor hydrogen and oxygen production. The chemical reaction at the core of H_2O_2 decomposition is expressed as $2H_2O_2 \rightarrow 2H_2O + O_2$. However, to guide the reaction toward hydrogen production, distinctive catalysts are integrated. An illustrative catalytic reaction facilitating hydrogen generation entails the utilization of catalytic blends or methodologies that facilitate the liberation of hydrogen from H_2O_2.

2.3.13 REDUCTION ELECTROLYSIS OF WATER (RED)

The Reduction Electrolysis of Water (RED) method is an innovative process that combines chemical reduction and electrolysis to produce hydrogen. This approach aims to enhance the efficiency of hydrogen production through electrolysis by reducing the energy required to separate water molecules into hydrogen and oxygen. The RED process involves two main steps: chemical reduction and electrolysis. In the chemical reduction step, water is mixed with a reducing agent, often a metal or a compound capable of donating electrons readily. This initiates a chemical reaction that partially reduces the water, forming hydrogen and oxides of the reducing agent. This reduction reduces the electrical energy needed for hydrogen production, as the chemical reaction supplies some of the energy required to break the H-O bonds in water. Subsequently, in the electrolysis step, the remaining compounds are separated, including the regeneration of the reducing agent to its initial state and the additional production of hydrogen. As a result, subsequent electrolysis can be performed with reduced energy expenditure compared to standard electrolysis because some water has already been converted into hydrogen during the chemical reduction step.

2.3.14 Carbon Dioxide Hydrogenation

The CO_2 hydrogenation method for hydrogen production, often referred to as the CO_2 conversion process into fuels and chemicals, leverages hydrogen to convert CO_2 into hydrogen-rich compounds like CH_4, methanol, and other hydrocarbons. While it doesn't directly yield hydrogen, this technique plays a pivotal role in CO_2 recycling and the generation of energy carriers from hydrogen. Hydrogen for this process is typically sourced from renewable origins, ensuring sustainability throughout. Operating on the principle of hydrogenation, this process demands a hydrogen input, often derived from renewable sources to ensure its eco-friendliness. The reaction occurs in the presence of specific catalysts, aimed at reducing CO_2 into hydrogenated products. Common reactions include the production of CH_4 through the Sabatier process, where CO_2 and hydrogen combine to form CH_4 and water, facilitated by a nickel catalyst. Additionally, methanol production involves reducing CO_2 into methanol, a versatile chemical and potential energy source, employing copper-based catalysts. Moreover, through intricate chemical reactions and advanced catalysts, CO_2 can be further converted into longer-chain hydrocarbons and alcohols, broadening the spectrum of available fuels and chemicals.

2.3.15 Water Photoelectrolysis

Photoelectrolysis of water represents an innovative approach to hydrogen production, harnessing solar energy to split water molecules into hydrogen and oxygen. Operating within a photoelectrochemical cell, this method utilizes light-sensitive semiconductor materials to facilitate the electrolysis reaction. Through the absorption of solar radiation, these materials generate electron-hole pairs, initiating the migration of electrons to the cathode and holes to the anode. This process enables the reduction of water at the cathode to produce hydrogen while oxidizing water at the anode to release oxygen. With its reliance on renewable solar energy and its ability to produce clean hydrogen without GHG emissions, photoelectrolysis offers a sustainable solution to the growing demand for hydrogen as an energy carrier. Despite its potential benefits, photoelectrolysis faces several challenges, including the need to improve its efficiency, enhance material durability, and reduce costs for widespread adoption. Efforts to enhance efficiency focus on developing semiconductor materials with better light absorption and charge transport properties, while also addressing losses during the conversion process. Additionally, research is underway to identify materials that can withstand the harsh conditions of intense illumination and aqueous environments to ensure long-term durability. Overcoming these challenges will be crucial in realizing the full potential of photoelectrolysis as a viable and sustainable method for hydrogen production, aligned with global efforts to transition to clean energy sources and reduce carbon emissions.

2.3.16 PHOTOBIOLOGY

Photobiology for hydrogen production is an innovative approach leveraging living organisms like algae and photosynthetic bacteria to convert solar energy into hydrogen, a clean energy source. This method harnesses the natural process of photosynthesis, where these organisms capture sunlight to transform CO_2 and water into carbohydrates and oxygen, with hydrogen generation as a potential by-product. Operating on the principles of water biophotolysis and photoassisted fermentation, photobiology offers several advantages. It relies on renewable solar energy, reducing reliance on fossil fuels and minimizing GHG emissions. Additionally, it can utilize nonfood biomass, mitigating concerns related to food security and land use. Moreover, the potential dual function of CO_2 sequestration and hydrogen production makes it an attractive option for addressing environmental challenges associated with carbon emissions. Despite these benefits, challenges such as conversion efficiency, stability of organisms, and hydrogen separation need to be addressed through ongoing research and development efforts to realize the full potential of photobiology for hydrogen production on a large scale.

2.3.17 HYDROGEN FROM INDUSTRIAL WASTE

Hydrogen production from industrial waste offers a promising avenue for both waste management and clean energy generation. Rooted in the principles of a circular economy, this method not only addresses the challenge of waste disposal but also contributes to reducing GHG emissions and advancing renewable energy. It encompasses a range of innovative processes, including gasification, pyrolysis, biomass steam reforming, anaerobic fermentation, and electrolysis, each tailored to efficiently convert different types of industrial waste into hydrogen. While this approach presents numerous advantages, such as waste reduction, renewable energy production, and GHG emissions reduction, it also faces several challenges. Optimizing efficiency and cost-effectiveness, ensuring the quality of the produced hydrogen, and managing the by-products generated during the conversion process are critical areas that require attention. Despite these challenges, continued research, technological advancements, and supportive policies can pave the way for waste-to-hydrogen conversion to become a significant contributor to the transition toward sustainable energy systems.

2.3.18 HYDROGEN PRODUCTION BY ELECTROCHEMISTRY

Hydrogen production through electrochemistry, commonly referred to as water electrolysis, is a method that employs electric current to split water molecules (H_2O) into hydrogen (H_2) and oxygen (O_2). This process unfolds within an electrolyzer, a device primarily comprising two electrodes (an anode and a cathode) submerged in an electrolyte solution facilitating the transfer of electric charge. Water electrolysis can be encapsulated by the following chemical reactions taking place respectively at the cathodic and anodic electrodes:

- At the cathode (reduction): $2H_2O(l) + 2e^- \rightarrow H_2(g) + 2OH^-(aq)$
  Under the influence of electric current, water is reduced to gaseous hydrogen at the cathode, releasing hydroxide ions (OH-).
- At the anode (oxidation): $2H_2O(l) \rightarrow O_2(g) + 4H+(aq) + 4e^-$
  Simultaneously, at the anode, water is oxidized to gaseous oxygen, liberating protons (H^+) and electrons. The hydroxide ions (OH-) produced at the cathode migrate to the anode, where they may either react with protons to form water again or contribute to the solution's conductivity. Likewise, the protons generated at the anode migrate to the cathode, facilitating the electrochemical cycle.

2.4 SUMMARY

This chapter provides a comprehensive exploration of various methods and technologies for hydrogen production, detailing their technical, economic, and environmental aspects as summarized in Figure 2.8. Traditional methods such as SMR, electrolysis, and coal gasification are thoroughly examined alongside emerging technologies like biomass conversion, thermochemical water splitting, and HTE using nuclear energy. The intricacies and implications of each method have been highlighted, making a strong case for the critical role of hydrogen in the global energy transition.

SMR is identified as the most prevalent method due to its high efficiency, although it produces significant CO_2 emissions unless paired with CCS technologies. Electrolysis, especially when powered by renewable energy sources, emerges as a cleaner method but faces challenges related to efficiency and cost compared to SMR. Coal gasification is noted for its high emissions, which makes it less favorable without effective CCS integration. These traditional methods provide a foundation for hydrogen production but also underscore the urgent need for cleaner, more sustainable technologies. Emerging technologies such as biomass conversion offer a promising sustainable alternative by converting organic materials into hydrogen. However, this method faces challenges in energy efficiency and technology costs. Thermochemical water splitting, which utilizes high temperatures typically provided by nuclear reactors, achieves high efficiency with low emissions, presenting another viable pathway for hydrogen production. HTE further leverages nuclear energy to enhance efficiency and reduce overall energy consumption, marking significant progress in hydrogen production technologies.

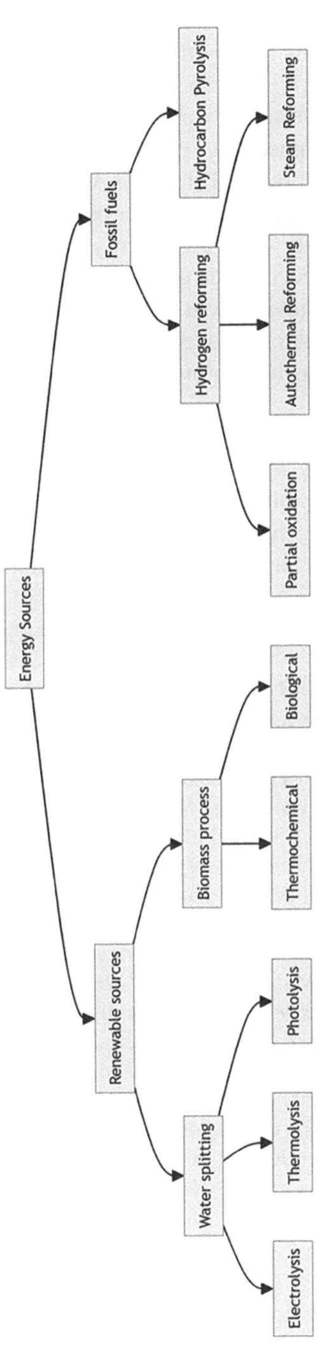

FIGURE 2.8 Methods of hydrogen production. (Re-printed with permission from ref. [9].)

The technical and economic challenges associated with hydrogen production have also been presented. It emphasizes the necessity for advancements in catalyst materials, membrane technologies, and system designs to improve the efficiency of electrolysis and other hydrogen production methods. Economic barriers are highlighted, particularly the high costs of setting up production infrastructure and the substantial investments required for distribution networks. Overcoming these challenges is crucial for making hydrogen a competitive energy carrier. Environmental and policy impacts are critically assessed through Lifecycle Assessment, which evaluates the environmental implications of different hydrogen production methods. Key metrics such as global warming potential, water use, energy use, and resource depletion are analyzed to provide a holistic view of hydrogen's environmental footprint. The chapter underscores the vital role of government policies in promoting hydrogen production. Financial incentives, regulatory standards, and international cooperation are identified as essential components in creating a supportive environment for hydrogen technologies.

The future perspectives emphasize the importance of stakeholder engagement in driving innovation and fostering a collaborative approach to advancing hydrogen production. By addressing technical, economic, and environmental challenges through concerted efforts, hydrogen can become a pivotal element in the global transition toward a cleaner and more sustainable energy system.

REFERENCES

1. Gao, F.-Y., P.-C. Yu, and M.-R. Gao, *Seawater electrolysis technologies for green hydrogen production: challenges and opportunities.* Current Opinion in Chemical Engineering, 2022. **36**: p. 100827.
2. Aldosari, O.F., I. Hussain, and Z. Malaibari, *Emerging trends of electrocatalytic technologies for renewable hydrogen energy from seawater: Recent advances, challenges, and techno-feasible assessment.* Journal of Energy Chemistry, 2023. **80**: p. 658–688.
3. Qureshi, F., et al., *Sustainable and energy efficient hydrogen production via glycerol reforming techniques: A review.* International Journal of Hydrogen Energy, 2022. **47**(98): p. 41397–41420.
4. Capurso, T., et al., *Perspective of the role of hydrogen in the 21st century energy transition.* Energy Conversion and Management, 2022. **251**: p. 114898.
5. Lee, S., et al., *Scenario-based techno-economic analysis of steam methane reforming process for hydrogen production.* Applied Sciences, 2021. **11**(13): p. 6021.
6. Kumar, S.S. and H. Lim, *An overview of water electrolysis technologies for green hydrogen production.* Energy Reports, 2022. **8**: p. 13793–13813.
7. Pal, A., et al., *Powering squarely into the future: A strategic analysis of hydrogen energy in QUAD nations.* International Journal of Hydrogen Energy, 2023. **49**: p. 16–41.
8. Jahanbakhsh, A., et al., *Underground hydrogen storage: A UK perspective.* Renewable and Sustainable Energy Reviews, 2024. **189**: p. 114001.
9. Arsad, A., et al., *Hydrogen electrolyser technologies and their modelling for sustainable energy production: A comprehensive review and suggestions.* International Journal of Hydrogen Energy, 2023. **48**(72): p. 27841–27871.

3 Hydrogen Storage

3.1 INTRODUCTION: GENERAL CONTEXT OF HYDROGEN STORAGE

Hydrogen storage is a critical element in the development and implementation of sustainable energy systems, playing a central role in the transition to a low-carbon economy. As an energy carrier, hydrogen offers particularly advantageous properties to meet contemporary ecological and technical demands. Its ability to release a substantial amount of energy per unit mass sets it apart from other energy sources, while its combustion, which only results in water, highlights its potential for significantly reducing pollutant emissions. However, the widespread adoption of hydrogen is hindered by storage challenges. Its physicochemical properties, especially its very low volumetric density, necessitate innovative storage solutions to effectively contain a sufficient quantity within a manageable volume. This involves the use of compression or liquefaction techniques, each presenting its own technical challenges and associated costs. Despite these obstacles, the transformative potential of hydrogen in the global energy landscape motivates ongoing research and the development of more efficient and less-costly storage technologies. This technical challenge only underscores the importance of hydrogen storage as a keystone for fully realizing the environmental and energy benefits of this revolutionary energy carrier.

To maximize the volumetric energy density of hydrogen, specific storage methods are required. High-pressure compression, typically ranging between 350 and 700 bars, is a common method but poses challenges in terms of safety and infrastructure costs. Similarly, liquefying hydrogen at cryogenic temperatures, approximately $-253°C$, achieves higher volumetric compactness but incurs high energy and financial costs due to the need for specialized equipment and the energy consumption for cooling. In response to these constraints, research has turned toward alternatives such as solid-state storage. This method utilizes advanced materials, such as metal hydrides, MOFs, or carbon nanotubes (CNT), which physically or chemically absorb hydrogen into their structures. Although these technologies are still in development, they promise significant reductions in the costs and risks associated with high-pressure or extreme-temperature storage. Nevertheless, challenges such as improving charging and discharging rates, as well as the longevity of the materials used, must be overcome for these solutions to become viable on a large scale.

The role of hydrogen storage in managing the inherent fluctuations of renewable energy sources is crucial for maintaining the balance between energy supply and demand. Hydrogen acts as an energy reservoir, allowing the capture of excess production during solar or wind energy peaks, which can then be stored for later use during periods of increased demand or reduced production. This storage strategy provides a flexible response to the challenges posed by the intermittency of renewable

DOI: 10.1201/9781032718453-3

energy, thereby improving the reliability and resilience of energy networks. Furthermore, it enables energy systems to maximize the use of renewable sources, reducing the need to resort to less sustainable energy sources during periods of energy deficit, and significantly contributes to the stabilization of the global energy grid.

To effectively integrate hydrogen as a major component of our energy systems, significant advancements in storage technologies are necessary. This involves not only improving the efficiency of current methods but also reducing the associated costs, making hydrogen storage competitive with other forms of energy. Current research initiatives are striving to develop innovative materials and improved processes that could revolutionize the way hydrogen is stored, thereby facilitating its large-scale adoption. Additionally, the use of hydrogen produced from renewable sources underscores its potential to minimize dependence on fossil fuels and reduce harmful emissions, aligning energy goals with global climate imperatives. This approach aims not only to optimize the economic viability of hydrogen but also to maximize its ecological benefits, contributing to a cleaner and more sustainable energy transition.

Hydrogen storage technologies vary significantly based on the storage scale, catering to different applications and needs as illustrated in Figure 3.1. For small, laboratory-scale storage, hydrogen is often stored in high-pressure gas cylinders, typically at 350–700 bar, which is straightforward and convenient for experiments. Another method for small-scale storage involves metal hydrides, materials that absorb hydrogen and release it when needed, providing a safe and compact solution. For medium, vehicle-scale storage, compressed hydrogen tanks are widely used in fuel cell vehicles, storing hydrogen at pressures up to 700 bar. This method balances storage capacity, safety, and weight considerations for mobile applications. Liquid hydrogen, stored as a cryogenic liquid at −253°C, offers a higher energy density than compressed gas, making it suitable for vehicles requiring extended range, such as buses and trucks. For very large-scale geological storage, salt caverns are commonly used, offering vast storage capacity and natural airtightness, ideal for bulk storage and seasonal energy management. Depleted oil and gas fields provide a cost-effective solution for massive hydrogen storage necessary for grid balancing and long-term energy security, while aquifers, subsurface porous rock formations saturated with water, offer an alternative large-scale storage option leveraging existing geological structures. These technologies enable the versatile use of hydrogen across various scales, supporting applications from research and transportation to large-scale energy systems.

In the following sections, different methods of hydrogen storage have been presented by exploring in detail the various types of tanks, the materials employed in their fabrication, and the technological processes. The advantages and limitations of each method, including high-pressure storage, low-temperature liquefaction, and emerging solid-state storage technologies, are analyzed. The focus is on comparing high-pressure tanks, cryogenic tanks, and materials such as metal hydrides, MOFs, and CNT, considering their efficiency, cost, safety, and durability. The latest innovations in tank design and fabrication are also examined, including the use of new alloys and composites that promise to improve performance and safety. This chapter aims to provide a comprehensive understanding of the technical challenges and future prospects of hydrogen storage, crucial for its viability as an energy carrier in a low-carbon economy.

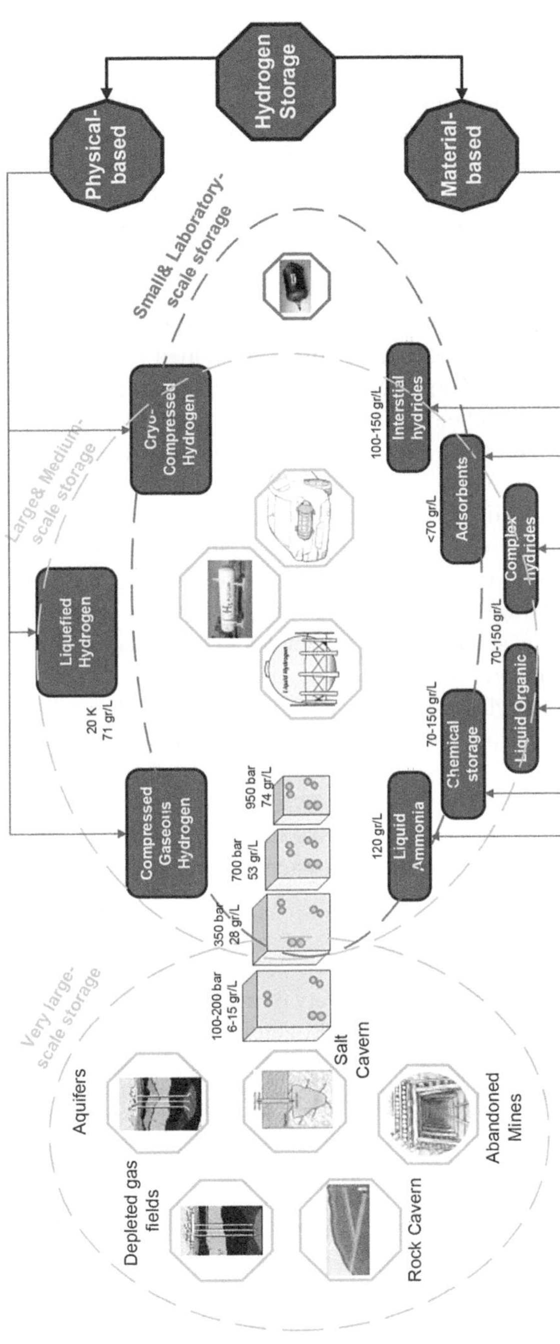

FIGURE 3.1 Hydrogen storage technologies with respect to its storage scale. (Adapted from ref. [1].)

3.2 CHEMICAL STORAGE OF HYDROGEN

Chemical hydrogen storage represents an advanced and strategic method for addressing the inherent challenges of storing this highly volatile gas. By forming stable chemical compounds, hydrogen is bound to other elements, eliminating the need to maintain it in its gaseous or liquid form at very low temperatures or under high pressure. This technique offers significant advantages, including improved safety through reduced risks of explosion and leakage, as well as increased energy density, allowing for a greater amount of hydrogen to be stored in a reduced volume.

Among the chemical hydrogen storage methods, metal hydrides are particularly notable. In this process, hydrogen is absorbed by specific metals or alloys, forming a solid by bonding with the host material. This interaction is generally reversible, allowing hydrogen to be released by heating the material, making the process suitable for repetitive applications. Metal hydrides are valued for their ability to store a large amount of hydrogen per unit weight, which is essential for mobile and portable applications. Additionally, chemical hydrides, such as ammonia (NH_3) and hydrazine (N_2H_4), offer an interesting alternative. These compounds, which contain hydrogen bonded to nitrogen or other elements, can be decomposed to release pure hydrogen. The decomposition can be induced thermally or catalytically, enabling efficient recovery of hydrogen. These substances are particularly useful for large-scale storage or in conditions where handling gaseous hydrogen would be impractical. These chemical storage technologies are crucial in fields requiring compact and secure hydrogen storage, such as fuel cell vehicles, where space and safety are major constraints. They are also relevant for storing energy from renewable sources, facilitating the conversion of electrical energy into chemically stored hydrogen during periods of excess production, and then reconverting it into electricity during periods of increased demand.

Ultimately, chemical hydrogen storage overcomes several challenges associated with more traditional forms of storage, paving the way for broader and more efficient use of hydrogen as an energy carrier in a low-carbon economy.

3.2.1 Physicochemical Phenomena

Chemical hydrogen storage involves a variety of complex physicochemical phenomena essential for the transformation of hydrogen into a storable form and its subsequent recovery. These phenomena are critical to the efficiency and safety of hydrogen storage systems. Key phenomena include:

3.2.1.1 Adsorption

Adsorption is the process by which atoms, ions, or molecules from a gas or liquid (adsorbate) adhere to the surface of a solid or liquid (adsorbent). In the context of hydrogen storage, this involves the attachment of hydrogen molecules to the surface of a material. Adsorption can be classified into two types: physical adsorption (physisorption) and chemical adsorption (chemisorption). Physisorption is driven by weak van der Waals forces and is typically reversible, making it suitable for applications requiring easy release of hydrogen. This process occurs at lower temperatures and is favored by high surface area materials such as activated carbons, MOFs, and CNT. These materials maximize the amount of hydrogen that can be physisorbed due to their large surface areas and pore volumes.

Chemisorption, on the other hand, involves the formation of chemical bonds between the adsorbate (hydrogen) and the adsorbent, resulting in stronger interactions that are less easily reversible. This process is more specific and often requires higher temperatures to break the chemical bonds for hydrogen release. Materials designed for chemisorption include certain metal oxides, hydrides, and functionalized MOFs, which facilitate strong interactions with hydrogen molecules. The choice between physisorption and chemisorption depends on the specific requirements of the hydrogen storage application. Physisorption is ideal for conditions where rapid adsorption and desorption are needed, whereas chemisorption is beneficial for long-term storage due to its stronger hydrogen interactions. Ongoing research in material science aims to optimize these adsorption processes to enhance the efficiency, capacity, and safety of hydrogen storage systems.

3.2.1.2 Absorption

Absorption, as distinct from adsorption, involves the penetration of hydrogen into the bulk volume of a material known as the absorbent. This phenomenon can occur in two main ways: either physically, where hydrogen simply dissolves in the material without altering its chemical structure, or chemically, where hydrogen reacts with the material to form a new chemical compound. In the context of chemical hydrogen storage, the formation of metal hydrides is a prime example of chemical absorption. Here, hydrogen is not only absorbed but also chemically bonded to the metal atoms, forming metal hydrides. Metals such as nickel (Ni), palladium (Pd), magnesium, and their alloys are commonly used for this type of storage due to their ability to form stable and reversible hydrides. The absorption reaction of hydrogen by these metals is exothermic, meaning it releases heat. When hydrogen is absorbed and metal hydrides are formed, the host metal's structure can expand due to the insertion of hydrogen into its crystalline lattice. This insertion often requires specific temperature and pressure conditions to optimize the metal's capacity to store hydrogen.

The process of hydrogen release, on the other hand, is endothermic, requiring the input of heat to break the chemical bonds formed between hydrogen and the metal, thereby allowing hydrogen to escape as a gas. This release process can be controlled by adjusting the temperature, which is crucial for applications such as energy generators where hydrogen must be released in a controlled manner to power a fuel cell. The efficiency of hydrogen absorption and release greatly depends on the nature of the metal used, the structure of the formed hydride, and the thermal management of the process. Optimizing these parameters is essential to maximize storage capacity, hydrogen charge and discharge rates, and the durability of the storage system.

3.2.1.3 Desorption or Dehydration

Desorption is the process by which hydrogen is released from the surface or volume of the material where it has been stored, either through adsorption or absorption. This phenomenon is crucial for the recovery of hydrogen for use in various applications, including fuel cells and energy production. The method by which desorption is induced closely depends on the initial storage type of hydrogen. For systems where hydrogen is physically adsorbed onto the surface of a material, often through weak interactions such as van der Waals forces, it can generally be released by simply modifying pressure or temperature conditions. For example, a decrease in pressure or an increase in temperature will reduce the interaction forces between hydrogen and the surface, allowing hydrogen to escape.

In cases where hydrogen is chemically absorbed, such as in metal hydrides, the desorption process is more complex. Here, hydrogen is not only dissolved but also chemically bound to the material, often a metal or alloy. To induce desorption, a considerable input of energy is often required to break the strong chemical bonds between hydrogen and the metal. This energy is typically provided in the form of heat. The application of heat increases the internal energy of the hydrides, gradually weakening the chemical bonds and allowing hydrogen to be released. The hydrogen release process can also be facilitated by the presence of catalysts that lower the activation energy required for desorption. These catalysts can be particularly useful in systems where rapid and efficient reversibility of adsorption and desorption is needed. It is also important to note that, in some cases, chemical changes can be utilized to induce desorption. For example, the introduction of another gas that reacts with the adsorbent material can release hydrogen through chemical displacement. In summary, desorption is a critical process that must be carefully managed to ensure effective and controlled release of hydrogen, thus enabling its optimal use in energy applications. Understanding and mastering this process are essential for the development of more efficient and safer hydrogen storage technologies.

3.2.1.4 Hydride Formation Reactions

The formation reactions of hydrides during chemical hydrogen storage are essential processes for converting gaseous hydrogen into a solid form, making it easier to store and handle. This phenomenon involves specific interactions between hydrogen and various metals, resulting in the creation of compounds called metal hydrides. The process initiates when gaseous hydrogen comes into contact with the surface of a metal. Metals used for hydride formation, such as magnesium, Ni, Pd, and titanium (Ti), possess catalytic properties that facilitate the dissociation of hydrogen molecules (H_2) into individual hydrogen atoms. This dissociation is often aided by the presence of defects or active sites on the metal surface.

Once dissociated into atoms, hydrogen begins to diffuse into the metal. This diffusion can occur either on the surface or deeper into the metal's crystalline lattice. The metal's ability to absorb hydrogen depends on its crystalline structure, porosity, and chemical composition. Diffusion is typically favored at higher temperatures, where thermal agitation enables hydrogen atoms to migrate more easily through the metal lattice. As hydrogen atoms infiltrate the metal's crystalline lattice, they bond with metal atoms to form a new metal hydride. This bonding results in a modification of the metal's crystalline structure. For example, in the case of magnesium hydride (MgH_2), each magnesium atom in the lattice bonds with two hydrogen atoms. This structural transformation is often accompanied by an expansion of the crystalline lattice, which can induce mechanical constraints in the material. Hydride formation is an exothermic reaction, meaning it releases heat. The reversibility of this reaction is a key characteristic of hydrogen storage, as it allows for the recovery of stored hydrogen by applying heat to induce an endothermic desorption reaction. The hydrogen charge–discharge cycle in a metal hydride can be repeated several times, although performance may decline over time due to material fatigue and other degradation factors.

In some cases, catalysts are added to enhance the efficiency of hydride formation and decomposition reactions. These catalysts can lower the temperature required for desorption and accelerate the hydride formation reaction, making the process more efficient and economically viable. In summary, hydride formation reactions lie at the heart of chemical hydrogen storage technology. They enable the safe and efficient handling of hydrogen, a key element for its utilization in various energy and industrial applications.

3.2.1.5 Catalysis

Many chemical hydrogen storage systems rely on catalysts to expedite the formation and decomposition reactions of hydrides. Catalysts decrease the energy barrier required for these transformations, thereby enhancing reaction efficiency and kinetics. The catalytic mechanism by which these catalysts facilitate hydride formation and decomposition reactions in chemical hydrogen storage is both intriguing and essential for process effectiveness. The process begins with reactants, primarily hydrogen in hydride formation, being adsorbed onto the catalyst's active surface. This step involves hydrogen molecule binding to the catalyst surface, facilitating hydrogen molecule (H_2) dissociation into hydrogen atoms (H). Under the catalyst's influence, the H–H bond in hydrogen molecules weakens and eventually breaks, generating free hydrogen atoms. This dissociation step is crucial as hydrogen dissociation often limits the reaction kinetics without a catalyst.

Dissociated hydrogen atoms diffuse across the catalyst surface and are eventually absorbed by the metal forming the hydride. In hydride decomposition, hydrogen migrates to the catalyst surface where it recombines into hydrogen molecules. The reformed hydrogen molecules then depart from the catalyst surface, freeing it for other reactions. Commonly used catalysts include Pd and Ni, both effective for hydrogen dissociation and often used in metal hydride storage systems. Ti and zirconium (Zr) are added as dopants in some metal hydrides to enhance their hydrogen storage properties by modifying the metal's crystalline structure and improving hydrogen diffusion. Despite being more costly, platinum is valued for its high catalytic efficiency, especially in low-temperature conditions and for reactions requiring high purity. These catalysts play a vital role in reducing the energy barrier of hydride formation and decomposition reactions, significantly improving hydrogen storage process kinetics and overall efficiency, thereby promoting wider adoption of this clean energy technology.

3.2.1.6 Diffusion Phenomena

The diffusion of hydrogen within materials is a vital process in hydrogen storage, occurring either at the material's surface or within its volume. This diffusion can manifest through volume mechanisms, which involve hydrogen traversing interatomic spaces within the crystal lattice, or via surface pathways, navigating along grain boundaries or pore surfaces. The efficiency of hydrogen storage and recovery is intricately linked to the material's atomic structure and surface chemistry, as these factors dictate the available diffusion pathways and rates.

Moreover, the behavior of hydrogen diffusion is heavily influenced by the morphology and texture of the material's microstructure. The presence of pores or interstices within this microstructure significantly affects hydrogen diffusion capacity, providing channels for hydrogen movement. In materials such as metal hydrides or carbon-based adsorbents, the size and connectivity of these pores are crucial for facilitating efficient hydrogen transport. Additionally, crystal defects and dislocations within materials create preferential pathways for hydrogen migration, impacting diffusion rates depending on their type and distribution throughout the material. Understanding the interplay between material microstructure and hydrogen diffusion is essential for optimizing storage systems and enhancing their practical efficiency in various applications.

Hydrogen diffusion within materials adheres to Fick's laws. The first law of Fick is harnessed to elucidate the flux of hydrogen across the material concerning the concentration gradient according to Equation (3.1):

$$J = -D \frac{\partial C}{\partial x} \tag{3.1}$$

where J represents the diffusion flux, D signifies the diffusion coefficient, C denotes the hydrogen concentration, and x denotes the position along the material.

The second law of Fick (Equation 3.2), tailored for nonstationary scenarios, delineates the temporal alteration of hydrogen concentration within the material:

$$\frac{\partial C}{\partial t} = \frac{\partial^2 C}{\partial x^2} \tag{3.2}$$

3.2.1.7 Phase Changes and Structural Transitions

When materials, notably metal hydrides, absorb hydrogen, they can undergo significant phase changes or structural transitions. Understanding these transformations is crucial as they directly influence the materials' ability to store and release hydrogen efficiently and repeatably. When hydrogen is absorbed by a material, it can cause an expansion of its crystalline lattice and lead to a phase change. A specific example is Ti, which can absorb hydrogen and transition from a compact hexagonal phase to a body-centered cubic phase, thereby altering its mechanical and chemical properties. These transformations can be reversible or irreversible. In the case of metal hydrides, although the structure may initially revert to its original form after hydrogen desorption, repeated cycles of absorption–desorption may cause damage or permanent structural alterations that degrade the material's performance.

It is noteworthy that the repeated storage capacity of a material refers to its ability to absorb and release hydrogen multiple times without significant loss of its initial storage capacity, a critical criterion for applications such as hydrogen vehicles that require frequent charging and discharging cycles. These repeated cycles can induce material fatigue, cracking, or changes in microstructure, negatively impacting storage capacity. Moreover, storage efficiency refers to the amount of hydrogen that can be effectively stored and recovered compared to the theoretical amount predicted by the material's chemical composition. This efficiency is also influenced by reaction kinetics, that is, how quickly hydrogen can be absorbed and released. Changes in the material's crystalline structure may slow down these processes, thereby reducing storage efficiency.

Factors influencing phase changes and structural transitions of materials used for hydrogen storage mainly include temperature and pressure. Increasing temperature can facilitate these transformations by providing the energy needed to overcome energy barriers, while pressure directly affects the amount of hydrogen that can be absorbed by the material. Additionally, the chemical composition and prior thermal or mechanical treatment of the material can also affect its crystalline structure and diffusion properties. Together, these elements modulate the material's ability to undergo transformations under the influence of hydrogen, thereby affecting its efficiency and durability as a storage system.

3.2.1.8 Thermal Phenomena

The formation and decomposition reactions of metal hydrides are essential in hydrogen storage systems and are characterized by significant thermal phenomena. These reactions can be exothermic, such as the formation of magnesium hydride (MgH_2) where hydrogen is absorbed by a metal, releasing heat, which can increase the system's temperature and influence the safety and energy efficiency of storage. Conversely, the decomposition of hydrides, such as that of magnesium hydride to release hydrogen, is an endothermic reaction that requires an input of external heat, which can cool the material and necessitate heat management to maintain practical reaction rates. Effective management of these heat flows is therefore crucial for the efficiency and safety of hydrogen storage systems. This heat management in the formation and decomposition reactions of metal hydrides plays a fundamental role in the efficiency, safety, and durability of hydrogen storage systems.

First, temperature directly influences the kinetics of chemical reactions; thus, controlling heat helps maintain optimal reaction rates, facilitating efficient absorption and release of hydrogen. Second, it is crucial to control temperature spikes, especially with exothermic reactions, to avoid degradation of storage materials or uncontrolled reactions that could lead to safety incidents. Finally, effective heat management helps minimize thermal stresses on materials, thereby prolonging their lifespan and preserving their structural integrity, which is essential for storage system durability. Hydrogen storage systems often incorporate practical solutions for thermal management, including the use of heat exchangers and residual heat recovery. Heat exchangers, for example, are crucial in these systems as they help regulate temperature during the exothermic and endothermic formation and decomposition reactions of hydrides. These devices facilitate the removal or addition of necessary heat, thus maintaining a stable and secure environment. On the other hand, excess heat generated during hydride formation can be recovered and reused for other industrial processes or heating, thereby increasing the overall energy efficiency of the system. These approaches demonstrate how effective heat management can not only enhance the performance and safety of hydrogen storage systems but also contribute to a more sustainable use of energy resources.

3.2.2 DIFFERENT METHODS OF CHEMICAL STORAGE

3.2.2.1 Metal Hydrides

The method of hydrogen storage through metal hydrides is an advanced technique that utilizes the properties of certain metals or alloys to absorb and desorb hydrogen (Figure 3.2). This technique is crucial in the current context of seeking sustainable and secure energy solutions. Here is a detailed examination of this method, covering its history, principles, materials, storage capacity, industrial applications, and future prospects.

The use of metal hydrides for hydrogen storage dates back to the 1970s, a period marked by intensified research on alternative energies following oil crises. These efforts aimed to find effective and secure means of hydrogen storage for various applications, particularly in the energy and transportation sectors. The method of hydrogen storage through metal hydrides relies on specific chemical interactions between hydrogen and certain metals or alloys, forming what are called metal hydrides. To express in more detail, let's focus on the nature of the hydride bond, its formation, and its strength. In metal hydrides, hydrogen atoms are bound to metal atoms by ionic or covalent bonds, often with a metallic component. The exact nature of the bond depends on the metal or alloy used. Metals that form hydrides can be divided into two main categories:

i.   Ionic Hydrides: Typical of alkali and alkaline earth metals such as calcium and magnesium, where hydrogen forms negative ions (H^-) and the metal forms positive ions. These hydrides generally have a well-defined crystalline structure.

ii. Covalent or Metallic Hydrides: Formed with transition metals such as Ni and Pd. In these cases, hydrogen can be considered as forming a covalent bond or a metallic bond with the metal's crystal lattice. Hydrogen is often located in the interstices of the metal's crystalline lattice, slightly altering the structure without destroying it.

The hydrogen absorption reaction is exothermic. When gaseous hydrogen is introduced under pressure near the metal, it dissociates into hydrogen atoms that are then

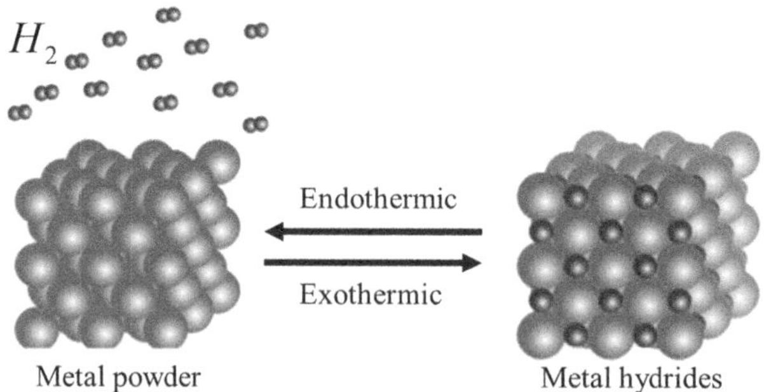

FIGURE 3.2   The principle of hydrogen storage through metal hydrides.

absorbed by the metal. These hydrogen atoms insert themselves into the interstices of the metal's crystalline lattice, forming the hydride. This insertion generally causes an expansion and distortion of the metal lattice. The strength of the hydride bond varies depending on the metal or alloy used. It is decisive for the temperature and pressure required for hydrogen release. In general, hydrides formed with alkali metals have stronger bonds and require higher temperatures to release hydrogen. Conversely, transition metals form hydrides with weaker bonds, allowing hydrogen to be released at lower temperatures, often more suitable for practical applications.

The bond strength directly influences the efficiency of hydrogen storage and release. Bonds that are too strong make hydrogen release difficult without significant heating, while bonds that are too weak may not store hydrogen stably enough for certain applications. The optimal balance allows for an efficient cycle of absorption and release, essential for applications such as fuel cells and renewable energy storage. A thorough understanding of these bonds and their manipulation through the choice of materials and operational conditions is therefore crucial for optimizing metal hydride technology for hydrogen storage. Materials used for storage in the form of metal hydrides include metals such as Ni, Ti, magnesium, and their alloys (Figure 3.3). Alloys are often preferred as they can be optimized to improve storage capacity and the speed of hydrogen absorption and release. Current research also focuses on nanotechnologies to enhance the performance of these materials.

The storage capacity of metal hydrides varies depending on the material used but can reach significant values, allowing for the storage of up to about 7.5% hydrogen by weight. This makes metal hydrides competitive compared to other hydrogen storage methods, including high-pressure tanks or liquid storage. Metal hydrides are used in several areas, including transport for FCEVs, where they offer a safer and more compact alternative to high-pressure tanks, renewable energy storage as a means of storing excess energy produced by renewable sources, and space and military applications where safety and compactness are crucial. The prospects for the use of metal hydrides are promising, especially with the development of renewable energy technologies and the need for efficient and secure energy storage systems. Continuous improvement of alloys, cost reduction, and increased efficiency of hydrogen release are key research areas that could further expand the use of this technology

FIGURE 3.3   Disk of magnesium hydride and storage of 25 kg of MgH_2.

(Ref: https://crd.ensosp.fr/doc_num.php?explnum_id=19144)

in the future. In conclusion, metal hydrides represent a highly promising hydrogen storage technology, offering a balance between energy density, safety, and potential application in various industrial sectors.

3.2.2.2 Liquid Organic Hydrogen Carriers (LOHCs)

The hydrogen storage method using organic hydride liquids, such as the transformation of toluene into methylcyclohexane, represents an innovative and efficient approach for hydrogen storage and transport. This method leverages reversible chemical reactions to store hydrogen in liquid form, offering an interesting alternative to more traditional methods. Here is a detailed overview of this method, covering its history, principles, storage capacity, chemical bonds involved, industrial applications, and future prospects.

Research on organic hydride liquids for hydrogen storage has gained popularity in recent decades, primarily driven by the search for efficient solutions for hydrogen transport and storage. Serious studies on systems like toluene/methylcyclohexane began in the 1990s, aiming to develop hydrogen storage systems that are both safe and economically viable. The process relies on the reversible chemical reaction between an aromatic hydrocarbon (like toluene) and hydrogen to form a saturated hydrocarbon (like methylcyclohexane). This transformation occurs under the influence of a catalyst at controlled temperatures and pressures. The hydrogen can then be released by dehydrogenation, where the saturated hydrocarbon is reconverted into its original aromatic form, releasing the stored hydrogen. To better understand hydrogen storage with organic hydride liquids, let's examine the toluene-methylcyclohexane system, which effectively illustrates the mechanisms of hydrogen storage and release via catalyzed reversible chemical reactions. In terms of reactions, toluene (C_7H_8) reacts with three molecules of hydrogen to form methylcyclohexane (C_7H_{14}) through hydrogenation:

$$\text{Toluene } (C_7H_8) + 3\ H_2 \rightarrow \text{Methylcyclohexane } (C_7H_{14})$$

Conversely, methylcyclohexane can be dehydrogenated to reform toluene and release hydrogen:

$$\text{Methylcyclohexane } (C_7H_{14}) \rightarrow \text{Toluene } (C_7H_8) + 3\ H_2$$

The catalyst, often Ni or platinum-based, plays a crucial role in activating the dissociation of hydrogen into free atoms, which then add to the carbon atoms of the benzene ring of toluene, gradually saturating the pi (π) bonds and converting the aromatic ring into a cyclohexane ring. During dehydrogenation, methylcyclohexane is heated under the influence of the catalyst, facilitating the breaking of the saturated C-H bonds and the release of hydrogen. The aromatic bonds of toluene are delocalized pi (π) bonds, stable and low-energy, requiring significant energy to break, whereas the aliphatic bonds of methylcyclohexane, formed after hydrogenation, are weaker sigma (σ) bonds and easier to break, facilitating hydrogen release. Generally, the pi (π) bonds in toluene are stronger than the sigma (σ) bonds in methylcyclohexane, meaning more energy is needed to hydrogenate toluene than to dehydrogenate methylcyclohexane, favoring the reversibility and control of hydrogen release reactions.

This storage system is used for applications requiring easy and safe hydrogen transport, such as hydrogen refueling stations and certain industrial processes where hydrogen is needed in large quantities at a specific location. Challenges include optimizing catalytic conditions, minimizing catalyst costs, and maximizing reaction reversibility to ensure a sustainable and economically viable system. Accordingly, the toluene-methylcyclohexane system offers a promising method for reversible hydrogen storage, with significant advantages in terms of safety and energy efficiency. The storage capacity of organic hydride liquids is generally competitive compared to other chemical hydrogen storage methods. The toluene/methylcyclohexane system can store approximately 6.2 wt% hydrogen, which is higher than many metal hydrides but lower than high-pressure or liquid hydrogen storage systems. In the case of toluene transformed into methylcyclohexane, the reaction involves the formation and breaking of C–H bonds and the conversion of aromatic bonds into saturated aliphatic bonds. These transformations are catalyzed and reversible, allowing the hydrocarbons to be reused for multiple hydrogen storage and release cycles.

Organic hydride liquids are particularly interesting for industries where hydrogen transport and distribution pose major logistical challenges, such as the energy industry or mobility applications requiring compact and secure hydrogen storage solutions. The use of organic hydride liquids for hydrogen storage is promising, especially in the current context of energy transition toward cleaner sources. Challenges to be addressed include reducing catalytic material costs, improving reaction cycle efficiency, and managing by-products. Additionally, developing suitable infrastructure for handling and distributing these liquids is crucial for their large-scale adoption.

3.2.2.3 Ammonia (NH_3)

Ammonia (NH_3) is increasingly valued as a promising vector for hydrogen storage and transport, primarily due to its advantageous physical and chemical properties. First synthesized on a large scale in the early 20th century through the Haber–Bosch process, initially for fertilizer production, NH_3 has gained importance as an alternative to fossil fuels for energy storage.

NH_3 is extremely promising due to its ability to store hydrogen densely and efficiently, with a capacity of 17.6% by weight, facilitating the reduction of costs and logistical challenges associated with its transport and storage. Unlike pure hydrogen, which requires extreme storage conditions, NH_3 can be liquefied under a moderate pressure of 10 bars at room temperature, simplifying storage and transport compared to gaseous hydrogen, which requires more extreme conditions. It can also be dissociated to release hydrogen and nitrogen through thermal or catalytic processes, a reaction useful for applications like fuel cells. Additionally, NH_3 can serve directly as a fuel in certain engines or gas turbines and presents significant environmental benefits, especially if produced from renewable sources, thereby reducing reliance on fossil fuels and CO_2 emissions. These properties make NH_3 an effective and practical solution for hydrogen storage, suitable for large-scale use in various industrial and energy sectors. NH_3 is produced by the exothermic reaction of nitrogen and hydrogen:

$$N_2(g) + 3H_2(g) \rightarrow 2NH_3(g)$$

It can be dissociated into these elements through endothermic reactions to release hydrogen:

$$2NH_3(g) \rightarrow N_2(g) + 3H_2(g)$$

With a high energy density, it is used in agriculture, energy production, and the chemical industry. Current research focuses on improving the energy efficiency of synthesis and dissociation processes, with developments toward more efficient catalysts and production methods using renewable energy. Given the growing interest in the hydrogen economy, NH_3 is being considered as a potential fuel for fuel cells and gas turbines, as well as for maritime and aviation transport, thus driving research and innovation to overcome the technical and economic challenges of its large-scale use.

3.2.2.4 Complex Hydrides

Complex hydrides, often composed of alloys or multi-metallic complexes, represent an advanced category of materials for hydrogen storage. This method exploits the ability of metals to form hydrides to store hydrogen densely and reversibly. Here is an in-depth exploration of this technology, including its history, specific examples, operating principles, involved chemical reactions, used materials, storage process, as well as current research trends and future prospects.

Complex hydrides began to be widely studied for hydrogen storage in the 1970s. The development of these materials was driven by the search for more efficient and safer hydrogen storage solutions than traditional methods such as high-pressure or very low-temperature storage. Among complex hydrides, the following are noteworthy: magnesium-nickel hydrides (Mg_2NiH_4) capable of storing about 3.6% hydrogen by weight, lanthanum-nickel alloys ($LaNi_5H_6$) known for their good hydrogen absorption and desorption properties, and Ti/irconium-based alloys used for their ability to operate at relatively low temperatures and their good cyclability.

Complex hydrides store hydrogen through a chemical reaction between a metal (or a metal alloy) and hydrogen to form a metal hydride. The hydrogen storage reaction can be generally represented by the following reaction:

$$Metal + H_2 \rightarrow MetalH_x$$

where x represents the number of hydrogen atoms bound to the metal lattice. The release of hydrogen occurs through the reverse reaction, which is often thermally activated:

$$MetalH_x \rightarrow Metal + H_2$$

The materials used for complex hydrides are generally metal alloys such as Ni, magnesium, lanthanum, Zr, and Ti, often combined to optimize hydrogen storage and release properties. These materials are chosen for their ability to form stable yet easily reversible hydrides at practical temperatures and pressures.

The storage capacity of complex hydrides is one of their most notable characteristics, distinguishing them from other forms of hydrogen storage. Complex hydrides stand out for their ability to store hydrogen with a high energy density, surpassing simple metal hydrides and high-pressure gaseous storage, making them ideal for applications where space and weight are limited, such as in vehicles or portable systems. These hydrides can store a significant amount of hydrogen, often expressed as a percentage of weight, with alloys like Mg_2NiH_4 storing up to 3.6% hydrogen and others potentially exceeding 7% under optimal conditions. A key advantage of complex hydrides is their reversibility, allowing multiple cycles of hydrogen absorption and release without significant loss of capacity, which is crucial for cyclic uses such as transportation or renewable energy storage systems. They are designed to operate at moderate temperatures and pressures, making their daily use more practical and reducing the energy required to control storage. Additionally, the alloy composition, including metals like Ti and Zr, plays an essential role in the storage properties of complex hydrides, as does the thermal and mechanical treatments that can alter the microscopic structure of the alloy, directly influencing the efficiency of hydrogen absorption and release.

Research in the field of complex hydrides focuses on improving storage capacity, hydrogen charge and discharge rates, and reducing material costs. Recent innovations include the addition of catalysts to enhance reaction kinetics and the discovery of new alloys that operate at lower temperatures and more moderate pressures. Complex hydrides are considered for a multitude of applications, including renewable energy storage, mobility systems such as hydrogen vehicles, and as a storage means for energy infrastructure. Future development will depend on overcoming technical challenges such as improving energy density, thermal management of storage systems, and cost reduction.

3.2.2.5 Borohydrides

The concept of using borohydrides for hydrogen storage dates back to the 1940s and 1950s, primarily for military applications where the need for portable, high-density energy sources was crucial. However, it is only in recent decades that this technology has gained attention for civilian applications, particularly due to advances in renewable energy technologies and electric vehicles. Borohydrides, such as sodium borohydride ($NaBH_4$), store hydrogen by forming stable chemical bonds with boron. During the hydrolysis reaction, these bonds are broken to release hydrogen gas. The typical reaction is:

$$NaBH_4 + 2H_2O \rightarrow 4H_2 + NaBO_2$$

This reaction is exothermic, releasing energy, but often requires a catalyst to occur at a practical temperature and rate for real-world applications. In borohydrides, hydrogen is bonded to boron through covalent bonds, which are relatively strong but less stable than bonds in water molecules. Boron itself has relatively low electronegativity, allowing these bonds to be broken under certain conditions to release hydrogen.

The hydrogen storage process using $NaBH_4$ typically involves the solid storage of $NaBH_4$, often in powder or granule form. To release hydrogen, $NaBH_4$ is brought into contact with water in the presence of a catalyst (usually metal-based like cobalt or

Ni), which helps to accelerate the hydrolysis reaction without being consumed. The storage capacity of borohydrides, particularly $NaBH_4$, is one of the main advantages of this hydrogen storage method. Borohydrides can store a significant amount of hydrogen per unit mass and volume, making them particularly attractive for applications requiring high energy density.

In fact, each mole of $NaBH_4$ (weighing about 37.83 g/mole) can theoretically produce 4 moles of hydrogen, weighing a total of about 8.04 g. This means $NaBH_4$ can release about 11.3% of its weight in hydrogen. In terms of hydrogen density, borohydrides also offer high values. $NaBH_4$ can store approximately 120 to 150 kg of hydrogen per cubic meter of solid material. This density is significantly higher than that of compressed gaseous hydrogen (about 40 kg/m^3 at 700 bars) or liquid hydrogen (about 70 kg/m^3 at $-253°C$). This makes borohydrides particularly useful for applications where space and weight are critical constraints, such as in FCEVs or portable technologies.

Although borohydrides offer a high theoretical hydrogen storage capacity, several practical challenges currently hinder their efficiency. First, the use of catalysts is necessary to trigger the hydrolysis reaction at practical temperatures and with an acceptable rate, which requires catalysts that are efficient, economical, and durable—a field of intensive research. Second, the regeneration of $NaBO_2$ back into $NaBH_4$ presents inefficiencies and costs that impact the overall economics of the storage process. Finally, controlling the hydrolysis reaction requires advanced systems to adjust hydrogen release according to the specific needs of different applications, further complicating the use of this technology.

Ongoing research aims to improve catalyst materials, optimize reaction conditions, and develop more efficient and cost-effective methods for borohydride regeneration. These innovations are crucial to increasing the practical viability of this hydrogen storage technology and realizing its full potential in terms of storage capacity. Borohydrides are being considered for several industrial applications, including fuel cells for electric vehicles, backup power systems, and portable devices where high energy density is needed. They are also being studied for space and military applications where portability and reliability are critical.

Current research focuses on improving catalyst efficiency and reducing costs associated with borohydride production. Efforts also include the regeneration of borates back into borohydrides, a crucial aspect of the economic viability of the process. Advances in nanotechnology offer new avenues for enhancing the reactivity and thermal management of hydrolysis reactions. The prospects for borohydride-based hydrogen storage are promising, especially with the increasing demand for clean technologies. Developing greener and more efficient processes for the synthesis and regeneration of $NaBH_4$ could make this technology more attractive and competitive compared to other forms of energy storage.

3.2.2.6 Carbides

The use of carbides for hydrogen storage is a relatively innovative approach in the field of energy storage. This method fits into the broader context of efforts to develop safe, efficient, and economically viable hydrogen storage solutions, essential for applications where hydrogen is used as an energy carrier. Here is a detailed explanation of this method, including its history, principles, chemical mechanism, storage capacity, industrial applications, examples of materials, as well as research and innovation prospects. Research on carbides for hydrogen storage is historically less documented than other methods, such as metal hydrides or high-pressure tanks. However, since the early 2000s, interest in developing new materials capable of more efficiently storing hydrogen has led to the study of carbides as potential storage materials. The basic principle of this method relies on the ability of carbides to react with hydrogen to form hydrides. Carbides, composed of carbon and another chemical element (usually a metal), can adsorb hydrogen under certain temperature and pressure conditions, integrating the hydrogen into their crystalline structure.

The chemical mechanism generally involves the reaction of carbides with hydrogen to form covalent or ionic bonds, resulting in the formation of hydrides. For example, titanium carbide (TiC) can absorb hydrogen to form a hydride, thus altering its crystalline structure. These reactions are often reversible, allowing not only the storage of hydrogen but also its release when needed. The storage capacity heavily depends on the type of carbide used as well as operational conditions such as temperature and pressure. Generally, the storage capacities of carbides are lower than those of traditional metal hydrides, but ongoing research aims to improve these performances.

Carbides for hydrogen storage can be used in various fields, including fuel cell vehicles, where controlled release of hydrogen is crucial, and renewable energy storage, where hydrogen can serve as a means to store intermittently produced energy from sources like solar or wind power. Some carbides being studied include TiC, tungsten carbide (WC), and silicon carbide (SiC), each with specific properties that influence their efficiency and viability as hydrogen storage materials.

Research continues to focus on improving storage capacity, reducing material costs, and optimizing operational conditions. Innovations in this field include the development of new carbide composites and alloys, as well as improvements in synthesis and processing methods to enhance their efficiency. The future of hydrogen storage via carbides appears promising, with potential advancements that could significantly improve their storage capacity and cost-effectiveness. The sustainable development of these technologies is crucial for their integration into future energy systems, particularly in the context of the energy transition to cleaner sources.

3.2.2.7 Hydrogen Storage in the Form of Formate

Storing hydrogen as formate represents an interesting and innovative approach, particularly for applications requiring controlled and clean hydrogen release at moderate temperatures. Here is a detailed explanation of this method, including its history, principles, chemical mechanisms, storage capacity, industrial applications, examples of materials, as well as research advancements and future perspectives. Interest in formate as a means of hydrogen storage began to emerge in the 2010s when researchers explored various chemical methods for hydrogen storage that could overcome the limitations of physical methods like compression or liquefaction. Formate stood out due to its stability and its ability to release hydrogen at relatively low temperatures.

Hydrogen storage in the form of formate involves using salts or esters of formic acid to store hydrogen. These chemical compounds can absorb or release hydrogen through reversible chemical reactions, making them potentially ideal for cyclical hydrogen storage. The key mechanism in using formates for hydrogen storage is the catalytic decomposition of formate to release hydrogen. This reaction is generally catalyzed by metals such as Pd or ruthenium. The reverse reaction can also be catalyzed to reform formate from hydrogen and CO_2 in some cases, allowing for the recovery of stored hydrogen. The chemical bonds involved are primarily covalent and ionic, depending on the specific structure of the formate used.

The storage capacity depends on the molecular density of formate and its efficiency in integrating and releasing hydrogen. Formates can potentially offer competitive hydrogen storage density, especially if reactions can be optimized to maximize hydrogen recovery. Formates could find applications in several industrial fields, including portable fuel cell systems where controlled hydrogen release is necessary, and stationary energy storage, allowing hydrogen produced by renewable sources during periods of overproduction to be stored for later use. Materials typically used for hydrogen storage in the form of formate include sodium formate salts (NaHCOO) and potassium formate salts (KHCOO), among others. Current research focuses on improving catalysts to make the decomposition and reformation of formates more efficient and economical. Optimizing reaction conditions and discovering new, more active catalysts are key areas of innovation. The prospects for hydrogen storage in the form of formate are promising, with possibilities for improving efficiency and reducing operational costs. The long-term viability of this method will depend on advancements in catalysis and the ability to integrate these systems into commercial and industrial applications.

3.2.2.8 Graphene and Carbon Nanotubes

Graphene and CNT are carbon-based materials that have captured the interest of the scientific and industrial communities due to their unique properties, especially for hydrogen storage. These nanotechnological materials are being studied for their potential to store hydrogen at high density through physical and chemical interactions with hydrogen. Here is a detailed exploration of these materials from various relevant aspects. Interest in the use of graphene and CNT for hydrogen storage increased significantly in the early 2000s. These materials were discovered and explored for various applications due to their exceptional properties, such as high specific surface area, high mechanical strength, and exceptional electrical and thermal conductivity.

Graphene, a single atomic layer of carbon arranged in a hexagonal lattice, and CNT, which are essentially rolled-up sheets of graphene, have the ability to store hydrogen through physical adsorption (physisorption) or chemical bonding (chemisorption). Physisorption takes advantage of the high specific surface area to accumulate hydrogen on the material's surface at relatively low pressures and temperatures, while chemisorption involves the formation of bonds between hydrogen atoms and carbon atoms.

In physisorption, the interactions are mainly van der Waals forces, which are relatively weak and reversible. In chemisorption, covalent bonds can form between hydrogen and defects or chemical functionalities on the surface of graphene or CNTs, leading to stronger but less easily reversible adsorption. Despite theoretically high storage capacities due to their large specific surface area, the actual hydrogen storage capacity of graphene and CNTs remains a subject of active research. Experimental results often show lower capacities than theoretical predictions, mainly due to material compaction and the difficulty of maintaining a surface accessible to hydrogen.

The potential application of graphene and CNTs in hydrogen storage could revolutionize sectors such as transportation, particularly FCEVs where high storage density is crucial, and renewable energy storage systems, where hydrogen can serve as a vector to store and release energy as needed.

Functionalized derivatives of graphene and CNTs modified with various functional groups to enhance their interaction with hydrogen are commonly studied to optimize storage capacity. Current research focuses on improving hydrogen storage and release capacities by exploring synthesis methods that allow better dispersion and greater surface accessibility of nanotubes and graphene. Additionally, the introduction of controlled defects or dopants into the structure of these materials to increase their chemisorption activity is an active area of innovation. Although promising, challenges related to production costs and manufacturing scale need to be overcome for graphene and CNTs to become viable solutions for large-scale hydrogen storage. Advances in the synthesis and manipulation techniques of these materials will be crucial to fully realize their potential in future industrial applications.

3.2.2.9 Comparison of Chemical Hydrogen Storage Methods

The different chemical hydrogen storage methods are compared in Table 3.1. This comparison is based on several essential criteria such as storage capacities, energy density, material density, and costs. This table clearly illustrates the diversity of available options for chemical hydrogen storage. The selection of the appropriate method will largely depend on the intended application, performance requirements, cost considerations, and existing infrastructure. It reflects current trends in the R&D of hydrogen storage technologies, highlighting the various trade-offs between storage capacity, energy density, material density, and costs. Each method has specific advantages and disadvantages that must be considered based on the target applications and operational requirements.

3.2.3 MATHEMATICAL MODELS OF CHEMICAL HYDROGEN STORAGE

3.2.3.1 Sorption Models of Metal Hydrides

These models describe the adsorption and desorption of hydrogen in metal hydrides based on thermodynamic and kinetic equations. They consider factors such as pressure, temperature, and alloy composition. The sorption models of metal hydrides for hydrogen storage are essential for understanding how hydrogen is absorbed and desorbed in metal-based materials. These models primarily focus on the thermodynamics and kinetics of the involved reactions. Several models are used to describe these processes, each with its own mathematical formulation. The main hydrogen sorption models are presented in this section.

TABLE 3.1

Comparison of Chemical Hydrogen Storage Methods

Storage Method	Storage Capacity (wt%)	Energy Density (MJ/kg)	Material Density (kg/m^3)	Costs
Complex Hydrides	3.6–7%	1.0–1.5	Variable depending on alloy	Quite high, active research for cost reduction
Borohydrides	10.8% (theoretical for NaBH4)	1.5–2.3	Approx. 1000–1200	High, complex regarding regeneration and catalysis
Carbides	Lower than hydrides	Variable	Variable	Research ongoing, potentially high due to complex synthesis processes
Formate	Competitive, specific to the material used	Comparable to hydrides	Variable	Developing research, potentially high due to the need for catalysts
Liquid Organic Hydrogen Carriers (LOHC)	Variable, depends on specific compound	0.6–0.8 (for toluene)	800–900 (for toluene)	Initially high, possible reduction with industrial optimization
Graphene and Carbon Nanotubes	Variable, often < 1%	Potentially high but not practically realized	Very light, unspecified details	Very high, currently limited by cost and production techniques
Hydrogen storage in formate	Variable, unspecified details	Variable	Variable	Ongoing research, potentially moderate to high depending on the production scale
Borohydrides	10.8% (theoretical for NaBH4)	1.5–2.3	Approx. 1000–1200	High, complex regarding regeneration and catalysis
Carbides	Lower than hydrides	Variable	Variable	Research ongoing, potentially high due to complex synthesis processes

3.2.3.1.1   Van't Hoff Model

The Van't Hoff model is used to describe the thermodynamic relationship between hydrogen pressure and temperature during sorption in metal hydrides. This model is fundamental for understanding how temperature affects the hydrogen pressure equilibrium in hydride materials. The mathematical expression of this model is presented according to Equation (3.3):

$$\ln P = \frac{\Delta H}{R} \cdot \frac{1}{T} + \Delta S \tag{3.3}$$

where P is the equilibrium pressure of hydrogen, ΔH is the enthalpy of hydride formation/dissociation, R is the ideal gas constant, T is the temperature (in Kelvin), and ΔS is the reaction entropy.

3.2.3.1.2   Sieverts Model

This model describes the solubility of hydrogen in metals, based on the assumption that the hydrogen pressure is proportional to the square root of the hydrogen concentration in the metal (Equation 3.4):

$$P = k\sqrt{C} \tag{3.4}$$

where P is the hydrogen pressure, C is the hydrogen concentration in the metal, and k is a proportionality constant.

3.2.3.1.3   Kinetic Sorption Model

These models describe the rate at which hydrogen is absorbed by metal hydrides. A particular case is the first-order chemical reaction model (Equation 3.5), where the sorption rate is proportional to the difference between the equilibrium hydrogen concentration and the instantaneous concentration:

$$\frac{dC}{dt} = k\left(C_{eq} - C\right) \tag{3.5}$$

where dC/dt is the rate of change of hydrogen concentration in the metal, C_{eq} is the equilibrium concentration, C is the instantaneous hydrogen concentration, and k is the reaction rate constant.

3.2.3.1.4   Jander Diffusion Model

This model is suitable for cases where hydrogen diffusion through the metal hydride is the limiting factor. It describes how the sorption rate decreases over time due to the decreasing concentration gradient (Equation 3.6):

$$\left(1 - \left(\frac{C}{C_{max}}\right)^{\frac{1}{3}}\right)^{2} = kt \tag{3.6}$$

where C is the hydrogen concentration at time t, C_{max} is the maximum hydrogen concentration, k is a constant related to the hydrogen diffusivity in the material, and t is time.

3.2.3.2　Kinetic Reaction Models

The kinetic reaction models for hydrogen storage describe how the rate of hydrogen storage and release is affected by various parameters such as temperature, pressure, and reactant concentration. These models are crucial for designing efficient and responsive hydrogen storage systems. Several kinetic models are used to analyze hydrogenation (storage) and dehydrogenation (release) reactions, and they are presented in the following.

3.2.3.2.1　Fractional Order Reaction models

These models assume that the reaction rate r, as a function of the reactant concentration (A), follows a noninteger order. The model can be expressed by the following Equation (3.7):

$$r = k\left(A\right)^{n} \tag{3.7}$$

where k is the rate constant and n is the reaction order, which is not necessarily an integer. A particular case of this model is when $n = 1$. In this case, the reaction rate will be proportional to the reactant concentration.

3.2.3.2.2　Langmuir–Hinshelwood model

This model is often used for catalytic reactions where two reactants are adsorbed onto the surface of a catalyst before reacting. If A and B are the reactants, the reaction rate can be expressed by Equation (3.8):

$$r = \frac{k\left(A\right)\left(B\right)}{1 + K_A\left(A\right) + K_B\left(B\right)} \tag{3.8}$$

where K_A and K_B are the adsorption constants of reactants A and B, respectively.

3.2.3.2.3　Eley–Rideal Model

In the Eley–Rideal model, one reactant is adsorbed on the catalyst while the other reacts directly with it from the gas phase. For a reaction between adsorbed A and gaseous B, the reaction rate is defined according to Equation (3.9):

$$r = k.\left(A\right)_{ads}.\left(B\right) \tag{3.9}$$

where $(A)_{ads}$ is the concentration of A adsorbed on the catalyst.

3.2.3.2.4　Temkin Model

The Temkin model takes into account the interactions between adsorbed molecules and the catalyst surface, suggesting that the adsorption energy decreases linearly with surface coverage. This model is less common for simple hydrogenation/dehydrogenation reactions but can be relevant for more complex systems.

3.2.3.3 Modeling Diffusion Processes

Models related to hydrogen diffusion processes are essential for understanding how hydrogen propagates through different materials, which is crucial in areas such as energy storage, hydrogen technologies, and advanced materials. Here are some of the most well-known models:

3.2.3.3.1 *Fick's Law*

The most classical approach to diffusion is based on Fick's laws. As discussed in section 3.2.1.6 (Diffusion Phenomena), the first Fick's law describes the diffusive flux J as proportional to the concentration gradient ∇C. The second Fick's law describes how the concentration changes with time t.

3.2.3.3.2 *Diffusion-Permeation model*

This model is often used to study the permeability of metals to hydrogen. It takes into account not only diffusion but also the absorption and desorption of hydrogen at interfaces. The basic equation can be expressed as follows, assuming a steady-state regime (Equation 3.10):

$$J = \frac{C_s - c_0}{e} \tag{3.10}$$

where C_s is the surface concentration, C_0 is the initial concentration, and e is the material thickness.

3.2.3.3.3 *Sieverts Model*

This model is particularly useful for describing the solubility of hydrogen in metals. Solubility coefficient, S, is directly proportional to the square root of hydrogen pressure P_H (Equation 3.11):

$$S = k \sqrt{P_H} \tag{3.11}$$

where k is a system-specific proportionality constant.

3.2.3.3.4 *Diffusion with Trapping model*

This model complicates Fick's laws by introducing trapping and detrapping terms of hydrogen, essential for understanding hydrogen behavior in materials where defects play a significant role. The general equation takes the form as shown in Equation (3.12):

$$J = -D\nabla C + k_{\text{trap}} \left(C_{\text{trapped}} - C \right) \tag{3.12}$$

where C_{trapped} represents the concentration of trapped hydrogen.

3.2.3.3.5   Models based on Quantum Mechanics

For analyses at the atomic scale, especially in complex materials or nanostructures, models using quantum mechanics or molecular dynamics are employed. These models are often specific to the system under study and may require solving the Schrödinger equation for systems of interacting electrons and nuclei.

3.2.3.3.6   Thermodynamic Models

These models allow for the calculation of changes in enthalpy and entropy associated with hydrogen storage and release reactions, providing insights into the feasibility and energy efficiency of the processes.

3.2.4   ANALYSIS OF TEMPERATURE AND PRESSURE EFFECTS ON CHEMICAL HYDROGEN STORAGE

Temperature and pressure play crucial roles in the efficiency and safety of chemical hydrogen storage. Variations in temperature can significantly affect the reactivity of the involved chemical reactions, directly impacting the materials' capacity to absorb or release hydrogen. For instance, some chemical compounds used for hydrogen storage require precise temperatures to maintain the balance between hydrogen absorption and release, which can influence the overall storage efficiency. Similarly, pressure affects the chemical storage capacity by altering the conditions under which these chemical reactions occur. High pressures can enhance hydrogen absorption in materials, while pressure reductions can facilitate its release. Accurately managing these parameters is therefore essential to optimize storage capacity, minimize material degradation risks, and reduce associated operational costs. Thus, the selection of materials and configurations for chemical storage must be tailored to meet the specific temperature and pressure requirements of the intended applications. Additionally, conducting specific life cycle assessments (LCAs) for the concerned technologies and materials is essential.

3.2.5   PERFORMANCE EVALUATION AND ENVIRONMENTAL IMPACT MANAGEMENT OF CHEMICAL HYDROGEN STORAGE SYSTEMS

To evaluate the performance and safety of chemical energy storage systems such as hydrogen storage, various measurement methods are employed. These methods include measuring both gravimetric and volumetric capacity and energy density, which are essential for applications where weight and space are critical. Sorption kinetics are also analyzed to determine the rate at which hydrogen can be absorbed and released, directly influencing the reactivity and efficiency of the material used. Pressure and temperature resistance tests are conducted to ensure the robustness and safety of the materials under extreme conditions. In addition to these direct measurements, assessing durability through repeated cycles of hydrogen loading and unloading is crucial for simulating long-term use. Energy efficiency is quantified by the energy balance, where the energy required for storage is compared to the energy recovered during hydrogen release, thereby identifying potential losses. Safety tests include leak tests and fire behavior assessments to anticipate the risks of fire or explosion, thus reinforcing preventive measures.

Alongside technical evaluation, managing environmental aspects is integral to the deployment of chemical hydrogen storage. This management begins with the choice of the hydrogen source, favoring green hydrogen produced via renewable methods to minimize the carbon footprint. Facility design aims to prevent leaks, while regular monitoring systems ensure the maintenance of environmental integrity. Energy efficiency is optimized to reduce indirect emissions, and the reuse and recycling of materials are encouraged to support a circular economy.

These approaches are crucial not only for minimizing the environmental impact of the production and use of storage materials but also for aligning industrial practices with regulatory environmental standards. Specific impacts, such as GHG emissions related to hydrogen production and water consumption in industrial processes, are evaluated through environmental impact studies before establishing new facilities. This multidimensional framework of technical measurements and environmental management ensures that hydrogen storage not only meets operational and safety requirements but also contributes to a sustainable energy transition.

3.2.6 CHALLENGES AND SOLUTIONS FOR EFFECTIVE DEPLOYMENT OF CHEMICAL HYDROGEN STORAGE

Chemical hydrogen storage, although promising for its energy applications, faces a series of technical, economic, and regulatory challenges that must be overcome to ensure its effective and safe large-scale deployment. The main constraints include managing the high pressures and low temperatures necessary for liquid storage, requiring robust tanks, efficient thermal insulation, and energy-efficient refrigeration systems. Safety challenges are also paramount, such as hydrogen embrittlement and material corrosion, which can compromise the structural integrity of the tanks. Additionally, exothermic reactions during the storage and handling of hydrogen must be rigorously controlled to avoid potential hazards like fires or explosions. Commercially, the barriers include the high costs of materials and manufacturing processes, as well as the expenses associated with installing and maintaining the necessary infrastructure. Chemical hydrogen storage often has lower energy efficiency compared to other storage methods, such as electric batteries, due to significant energy losses during conversion processes. The technical complexity of storing and safely handling hydrogen requires advanced technologies to manage high pressures and temperatures and to resist embrittlement and corrosion phenomena. Furthermore, developing a suitable distribution infrastructure is a major challenge, both costly and logistically complex, involving the construction of filling stations and secure transportation networks. Public perception may also hinder the adoption of chemical hydrogen storage due to safety concerns and a lack of awareness about its environmental benefits. Additionally, the absence of coherent regulatory frameworks and well-established standards for hydrogen storage and transport can discourage significant investments.

These aspects must be addressed through cost reduction, efficiency improvements, the establishment of clear regulatory standards, increased awareness of hydrogen's benefits, and ongoing innovation in storage technologies and sustainable hydrogen production. Continuous research is crucial to develop materials and alloys resistant to embrittlement and corrosion, improve surface treatments, and optimize engineering and design processes to maximize the durability and safety of storage systems.

3.2.7 Integration and Impact of Chemical Hydrogen Storage in the Global Energy Transition

Chemical hydrogen storage is an essential component in advancing renewable energy systems, facilitating a sustainable energy transition, and providing solutions to overcome the intermittency of sources such as solar and wind. This process not only allows for the storage of excess energy produced during peak production periods for later use but also serves as an energy carrier to reduce dependence on fossil fuels, thereby improving energy security and reducing carbon emissions. Hydrogen can be converted into electricity, used as fuel for transportation, or serve various industrial and residential applications, contributing to the decarbonization of multiple sectors. Optimal integration of chemical hydrogen storage into the overall energy system requires synergy with renewable energies, effective interconnection between energy sectors, and the development of appropriate infrastructure, such as hydrogen distribution networks and refueling stations. Additionally, it is crucial to establish regulatory frameworks to support this integration, including safety standards, environmental regulations, and financial incentives to promote green hydrogen.

Current regulations must evolve to address emerging challenges related to increasing hydrogen production and its integration into energy networks. This includes international standards for certifying hydrogen sustainability, specific regulations for large-scale storage, and standards for conversion technologies like fuel cells. Simultaneously, international collaborations play a pivotal role in developing chemical hydrogen storage by sharing knowledge, resources, and technologies. Initiatives like the Hydrogen Council, Mission Innovation, and various bilateral agreements and specific collaborative projects bring together countries, companies, and nongovernmental organizations to accelerate hydrogen adoption. These joint efforts are vital to overcoming technical and economic barriers and ensuring the seamless integration of hydrogen into global energy systems, making hydrogen a cornerstone of the low-carbon economy.

3.3 PHYSICAL STORAGE OF HYDROGEN

Physical hydrogen storage is essential for its effective use in various sectors, including fuel cell vehicles and renewable energy storage systems. The main methods include compressed storage, where hydrogen is compressed to very high pressures (350 to 700 bars) in composite material tanks, or at lower pressures requiring larger storage volumes. Cryogenic storage liquefies hydrogen at approximately −253°C, optimizing storage by reducing volume but requiring special thermal insulation to maintain the liquid state. Metal hydride storage captures hydrogen in metals forming a hydride, from which hydrogen can be released by heating, ideal for controlled release. Chemical storage involves compounds that store and release hydrogen through chemical reactions, and clathrate storage traps hydrogen in water molecule structures at temperatures close to 0°C. Each method has advantages and disadvantages regarding cost, energy density, safety, and ease of use, influencing their choice based on specific application needs.

Compressed hydrogen storage is a widely used technique, especially in the mobility and energy sectors, to meet compactness and efficiency requirements. High-pressure storage, often between 350 and 700 bars, is preferred in contexts where space is limited, such as in fuel cell vehicles. This type of storage allows a significant amount of hydrogen to be condensed in a small space, which is crucial for optimizing space and weight in vehicles. However, this method poses several industrial challenges, including the need for advanced technologies to ensure safety, the durability of the lightweight yet robust composite materials used, the longevity of tanks that must withstand repeated filling and emptying cycles, and the establishment of infrastructure capable of quickly recharging these tanks at high pressures.

On the other hand, low-pressure storage is less common and used where volume constraints are lower, such as in some stationary storage or large-scale industrial applications requiring large tanks. Although this method potentially reduces material costs, it demands significant investment in infrastructure to handle larger volumes of hydrogen and may be less suitable for applications requiring mobility or a small spatial footprint. The flexibility and scalability of this type of storage can, however, be advantageous for applications like storing excess wind or solar energy. In Table 3.2, the various physical hydrogen storage methods are compared in terms of storage quantities, costs, targeted industrial domains, main advantages, and challenges associated with each method.

High-pressure storage, especially at high pressure, is widely adopted in the automotive industry, particularly for fuel cell vehicles, as it allows storing a large amount of hydrogen in a small volume. The low-pressure version, less expensive but bulkier, is suitable for installations where available space is not a constraint. Cryogenic storage, favored for its high energy density, is essential in areas such as aerospace and certain types of maritime transport where compactness is crucial. Metal hydrides, which release hydrogen by heating, are ideal for applications requiring controlled hydrogen release but are generally expensive and heavy. Chemical storage, heavily dependent on the material used, can be very efficient but often costly and requires specific conditions for hydrogen release. As for clathrate storage, still in the experimental stage and mainly used for research, it offers a storage method under less stringent conditions than other techniques, although its efficiency and scalability for large-scale use still need to be proven. Each method presents unique advantages and challenges, and the choice of the appropriate method largely depends on the specific requirements of each application as well as the costs associated with the necessary technology and infrastructure.

Current innovations to improve the physical storage of hydrogen primarily aim to increase energy density, reduce costs, and enhance system safety. Among the most promising advancements is the development of new carbon fiber-based composites for high-pressure tanks, capable of withstanding high pressures and optimizing storage efficiency by increasing the amount of stored hydrogen. Another approach is

TABLE 3.2

Technical Comparative Table Regarding the Different Methods of Physical Storage of Hydrogen

Storage Method	Hydrogen Quantity (kg H_2/m³)	Costs	Industrial Sectors	Advantages	Challenges
High-pressure storage	20–40	High	Transport, automotive	Compact, rapid refueling	Material costs, safety, durability
Low-pressure storage	10–15	Moderate	Stationary storage	Lower costs, simplicity	Significant storage volumes
Cryogenic storage	70–80	Very high	Aerospace, transport	High energy density, compact	Thermal insulation, operating costs
Metal hydride storage	20–150	Very high	Renewable energy, transport	Safety, controlled hydrogen release	Cost, weight, slow reaction kinetics
Chemical storage	Variable, depends on material	High	Energy, heavy industry	Potential for high density, stability	Costs, dependence on temperatures and catalysts
Clathrate storage	0.9–1.5 (per liter of water)	Experimental	Research, niche applications	Low pressure, increased safety	Specific conditions, energy efficiency

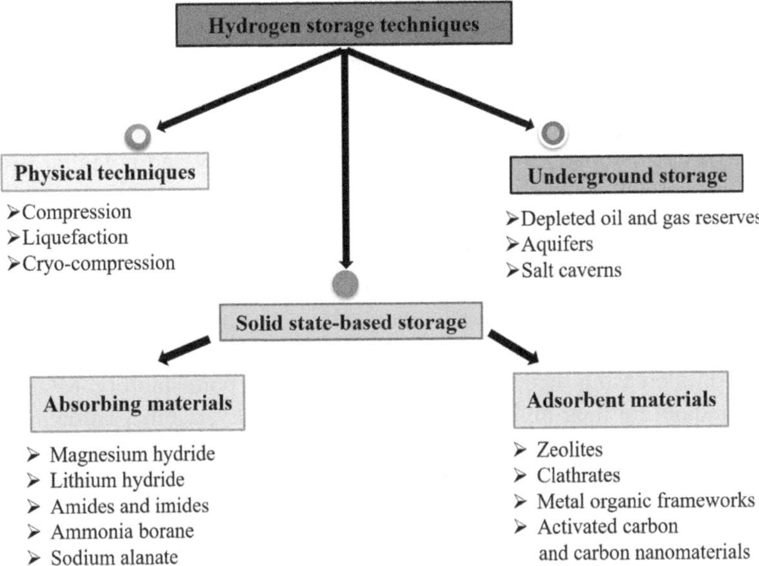

FIGURE 3.4 Classification of hydrogen storage technologies into three major categories.
(Adapted from ref. [2].)

cryogenic storage, involving storing hydrogen in liquid form at very low temperatures, with improvements in insulating materials and refrigeration technologies to minimize energy losses. Additionally, metal hydrides, capable of absorbing and releasing hydrogen under certain conditions, are being explored for their potential to operate at ambient temperatures with rapid cycles. Advanced porous materials, such as MOFs and boron hydrides, are also being studied for their ability to physically adsorb hydrogen at high density under moderate pressures. Nanotechnology is being explored to refine the structure of storage materials, allowing better management of hydrogen release and absorption. Finally, solid-state storage by chemical absorption is under development to use chemical compounds that react reversibly with hydrogen, optimizing pressure and temperature conditions for increased efficiency. These advancements aim to make hydrogen storage technologies safer and economically viable for widespread applications in the transportation, industry, and renewable energy sectors. Figure 3.4 presents a succinct summary of different hydrogen storage technologies.

3.3.1 Physical Adsorption on High Surface Area Materials

The method of physical adsorption, often referred to by the acronym PHA (Physical Hydrogen Adsorption), constitutes a promising technique for hydrogen storage, offering an efficient alternative to conventional methods of storage under pressure or in liquid form. This process relies on the adsorption of hydrogen molecules on the surface of porous materials at low temperatures, without any chemical change to the material or hydrogen. The materials used, such as activated carbons, molecular sieves, or MOFs, possess extremely vast internal surfaces that can trap a large quantity of hydrogen through relatively weak van der Waals forces.

This storage mode is particularly attractive due to its ability to operate at lower pressures and more moderate temperatures than those required for liquid or compressed hydrogen storage, thus providing increased safety and potential savings in terms of infrastructure and energy costs. Furthermore, physical adsorption allows for near-total recovery of the adsorbed hydrogen, which is essential for applications where the efficiency and repeatability of the storage cycle are crucial. However, the storage capacity and kinetics of hydrogen adsorption and desorption are strongly influenced by the nature of the adsorbent material. Current research is therefore focused on the development and improvement of adsorbent materials, with particular emphasis on increasing their specific surface area and pore stability. MOFs, in particular, stand out for their high structural and functional variability, allowing for specific customization to optimize interactions with hydrogen.

Challenges to overcome include improving the hydrogen storage density at ambient temperatures and under moderate pressures, and developing efficient regeneration processes that do not compromise the structural integrity of adsorbent materials over many charge and discharge cycles. Despite these obstacles, physical adsorption remains a highly active research avenue, with its advancements potentially revolutionizing hydrogen storage strategies for applications ranging from automotive to large-scale energy distribution networks.

3.3.1.1 Theoretical Aspects

The adsorption of a gas like hydrogen by a solid is the increase in the density of this gas at the surface of the solid due to the effect of intermolecular or interatomic forces. This phenomenon involves atoms, ions, or molecules, known as adsorbates, in a gaseous or liquid state forming physical or chemical bonds with the surface of a solid, called the adsorbent. This phenomenon has been known for a long time, primarily through the use of activated charcoal in medical applications or water purification. However, since the early 19th century, the capture of a gaseous species by a solid has been utilized. This phenomenon should not be confused with absorption, in which a material in a gaseous, liquid, or solid solution state enters the volume of another material in a liquid or solid state.

Adsorption can be purely a chemical phenomenon, known as chemisorption, or purely physical, known as physisorption. Indeed, depending on the binding energies involved, adsorption is divided into two types. Chemisorption is an irreversible phenomenon due to strong chemical bonds of covalent or ionic type between the atoms on the solid surface and the adsorbed molecules. The interaction forces involved are often short-range and can significantly alter or disturb the structures of the adsorbent and adsorbate. This type of adsorption involves high attraction energies, leading to high heats of adsorption, often approaching the energies of covalent chemical bonds in the range of 20 to 80 Kcal/mol. Physisorption occurs through the formation of physical bonds between the atoms on the solid surface and the adsorbed molecules. These bonds are the result of long-range van Der Waals attraction forces and short-range repulsion forces between the adsorbate and the adsorbent. Unlike chemisorption, physical adsorption often occurs at low temperatures or high pressure. The adsorbed molecule is fixed at a specific site and can move freely at the interface. The interactions between the adsorbent and adsorbate are mostly electrostatic, hence weak and reversible. Physisorption is rapid and does not alter the adsorbed molecules. It is characterized by low adsorption energy, in the order of 2 to 10 Kcal/mol.

Physical adsorption increases with the increase in gas pressure and is more significant at lower temperatures and with a larger interaction surface area with the solid. Adsorption decreases when the pressure decreases and/or the temperature increases. It allows for significant gas storage at low temperatures. Using this surface phenomenon to store gas requires a solid with a large interaction surface area, meaning a highly porous solid with a tortuous and highly rough surface. Most adsorption storage processes involve physisorption rather than chemisorption (Figure 3.5).

The adsorption capacity is often represented by isotherm curves, which can provide insight into the adsorption mechanism. The adsorption isotherm shows the variation of the amount adsorbed as a function of the equilibrium pressure at a constant temperature. The International Union of Pure and Applied Chemistry (IUPAC) classifies adsorption isotherms into six types (Table 3.3).

3.3.1.2 Adsorption Models

The phenomenon of adsorption is modeled by several models. The goal is to study the adsorption mechanism and estimate the absolute amount adsorbed. Among these models, some are simple and can be treated analytically, while others are more complex and require numerical methods.

The **Langmuir model**, proposed by Irving Langmuir in 1916, is a simple theoretical model with several assumptions: only a monolayer of adsorbate is formed, the adsorbent is a homogeneous system, and there is no interaction between adsorbate molecules. This model describes adsorption on an almost flat surface and considers that at equilibrium, the adsorption rate is equal to the desorption rate. The adsorbent's charge tends toward a maximum limit corresponding to the total occupation of the adsorption sites by a monolayer of the target substance. The **BET model**, proposed by Stephen Brunauer, Paul Emmett, and Edward Teller in 1938, models multilayer adsorption. It is designed to describe Type II adsorption isotherms observed in gases near their boiling points. This model is well-suited to express the adsorption

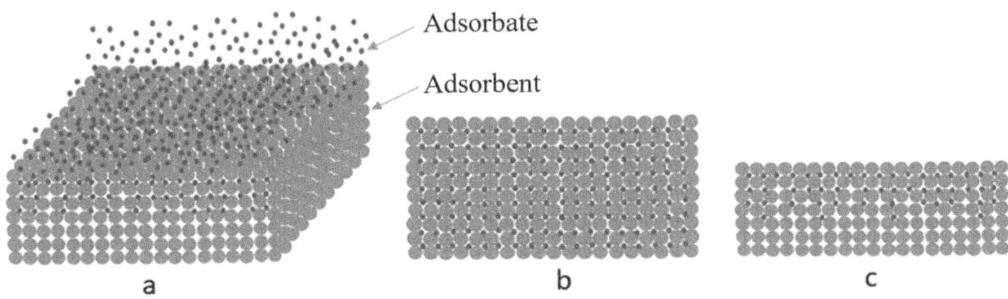

FIGURE 3.5 Schematic representation for adsorption of a gas by a solid: (a) adsorbent/adsorbate; (b) surface directly in contact with the gas; and (c) adsorption in the thickness.

TABLE 3.3

Description of Different Adsorption Models

Isotherm Type	Description	
Type I	Monolayer adsorption on microporous substrates Strong adsorbate–adsorbent interactions.	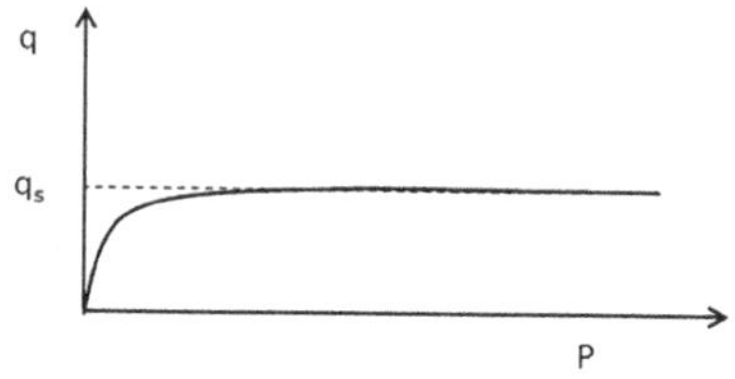
Type II	Monolayer adsorption at low pressure Multilayer adsorption at higher pressures Characteristic of nonporous solids	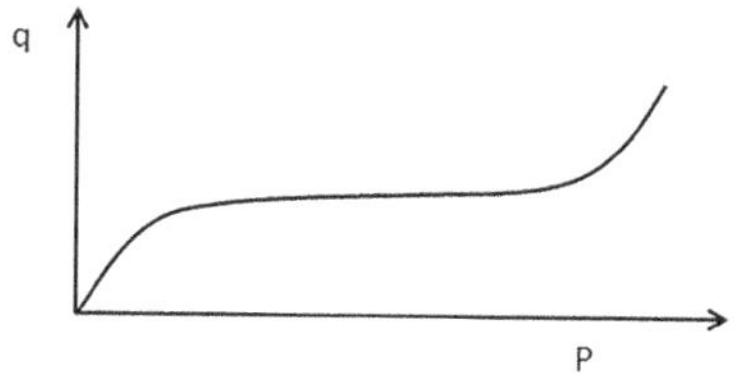
Type III	Exclusively multilayer adsorption Low adsorption capacity Weak adsorbate–adsorbent interactions	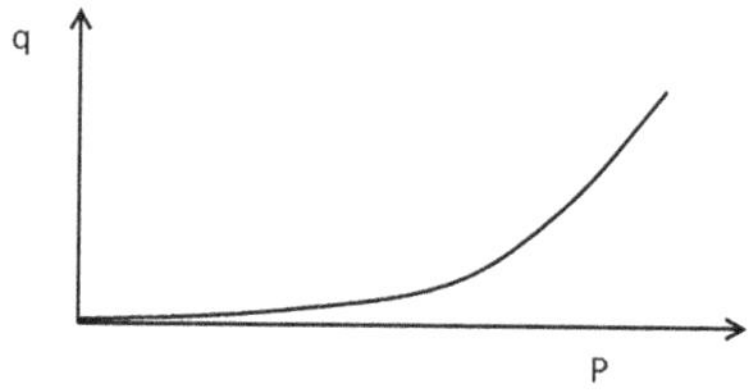
Type IV	Mesoporous adsorbents Resembles Type II at low pressures Shows a plateau due to capillary condensation Hysteresis present	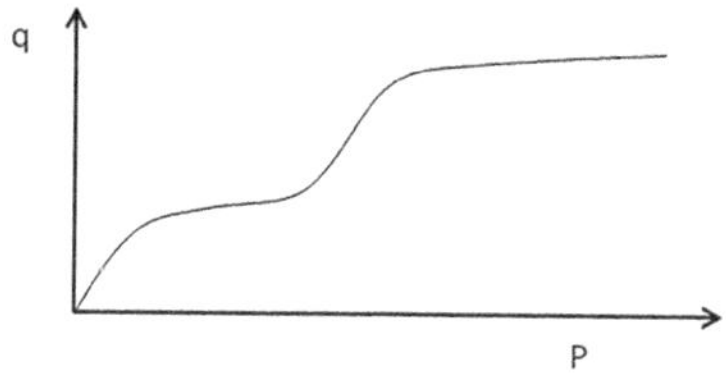
Type V	Similar to Type III at low pressures Resembles Type IV at high pressures Shows hysteresis	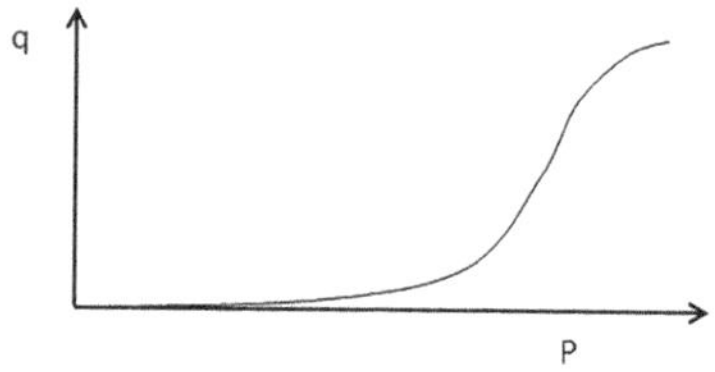
Type VI	Stepwise multilayer adsorption Typical for heterogeneous adsorbents	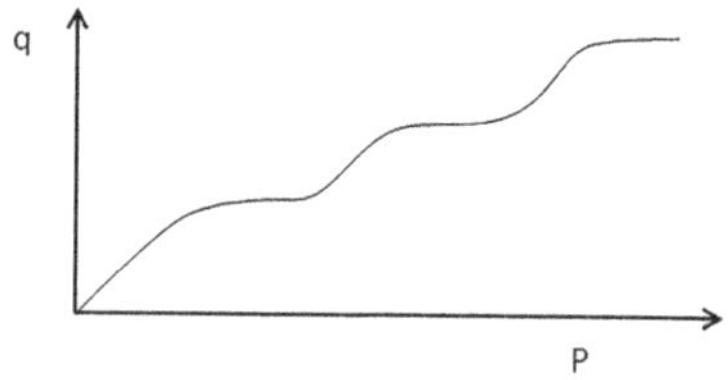

phenomenon, mainly of gases at high pressure on a solid adsorbent. It also describes adsorption on an almost flat surface and takes into account the saturation phenomena, involving the solubility of the solid in the solvent as the saturation concentration. The **Dubinin–Astakhov** model is proposed to describe the adsorption of a gas in the narrow pores of a microporous adsorbent. Originally, this model was proposed to describe the adsorption of subcritical gases. The model includes parameters like the micropores filling rate, the amount adsorbed, the maximum amount adsorbed, and a shape factor of the distribution, often assumed to be equal to 2 for activated carbons. The adsorption potential depends on temperature and pressure. The **Toth model** is an empirical model often used to describe Type 1 adsorption isotherms. It includes parameters such as the adsorption energy and the heterogeneity parameter of the adsorbent, generally less than 1. When the heterogeneity parameter is equal to 1, the **Toth** equation reduces to the Langmuir equation. The closer this parameter is to 1, the more homogeneous the adsorbate–adsorbent system. Generally, this parameter is assumed to be constant, but the Toth model adapts better to different isotherms when it is considered a function of temperature. The **Freundlich model** is an empirical relationship proposed by Herbert Freundlich in 1909. It allows representing both monolayer and multilayer adsorption at low adsorbate concentrations. This model is not based on any theoretical foundation, yet it is still widely used to represent many industrial applications. This model includes parameters such as the amount adsorbed per gram of solid, the equilibrium concentration of the adsorbate, and the Freundlich constants.

3.3.1.3 Types of Adsorbents

Adsorbents have a very large specific surface area (m^2/g), which does not appear on the exterior of the grains or particles of the material but is due to a very high porosity composed of a large number of very small pores. For example, activated carbon can have a porosity of 50 to 80%, consisting of pores 1 to 4 nm (10–3 μm) in size, providing a specific surface area of 500 to 1500 m^2/g. The specific surface area of the adsorbent is distributed over the walls of macropores (diameter > 50 nm), mesopores (diameter > 2 nm and < 50 nm), and micropores (diameter < 2 nm). Adsorption is a balanced phenomenon that can be represented by the equation: Free substance + Adsorbent $\rightleftarrows$ Adsorbed substance + Heat. For physical adsorption, especially for hydrogen storage, several types of materials are used because of their large specific surface area and their ability to store gas molecules at low temperatures.

3.3.1.3.1 Activated Carbons

At the end of the 19th century, activated carbon, in the modern sense of the term, appeared, although the first uses of activated charcoal, or "activated carbon," date back to the Egyptians. They utilized its adsorptive properties for medicinal purposes and to purify oils. However, it should be noted that at that time, it was mainly charcoal. The invention of activated carbon as we know it today is attributed to a Russian named Raphael Ostrejko, who sought to improve the adsorptive properties of charcoal for sugar decolorization in the early 1900s. Activated charcoals are the most manufactured and used industrial adsorbents, with various applications, particularly in water and air purification. The adsorption power of activated charcoals is primarily attributed to porosity, specific surface area, and surface functional groups.

The manufacture of activated carbon involves the use of various carbon-rich sources, including coals (anthracite, lignite, and bituminous) and woody plants such as wood and coconut shells. This process generally includes two steps: first, the low-temperature pyrolysis of the starting materials, followed by high-temperature carbon activation. This activation leads to the formation of a cavernous structure with large channels, resulting in high porosity and a large specific surface area. The two main activation techniques used are chemical activation and steam activation. Chemical activation is typically used for peat- or wood-based materials, involving strong dehydrating agents and high-temperature calcination, resulting in a very open structure ideal for adsorbing large molecules. Physical activation, used for charcoals and carbonized coconut shells, occurs at higher temperatures in the presence of steam, producing a fine-to-medium pore structure suitable for both liquid and gaseous adsorption.

Activated carbon is a term covering a variety of crystalline carbon-based adsorbents, distinguished by its highly developed internal porosity. To optimize their performance, it is essential to understand their physical and adsorption characteristics, mainly influenced by specific surface area and porosity. Initially, raw materials exhibit a graphitic carbon structure, with spaces between the crystals filled with disordered carbon or tars, which are eliminated during activation. After this stage, the crystals transform into carbon flakes with a thickness of about 1 to 2 nanometers and lengths varying from 2 to 60 nanometers. The specific surface area is higher when the crystallites are smaller, and determining the surface area of the lamellar carbon crystals can be done based on their geometric characteristics. Porosity and pore size distribution are crucial, with pores of various sizes forming a porous assembly after activation, modeled using cylindrical pores whose diameter can be determined by specific relations involving porous volume and surface area.

3.3.1.3.2 Zeolites

Zeolites are materials composed of aluminum, silicon, and oxygen, forming a three-dimensional aluminosilicate structure with a complex architecture of interconnected channels and cavities of molecular size, providing high porosity. This porous network allows for efficient physisorption where hydrogen molecules are attracted and held on the surface by van der Waals forces, a process that does not alter either the hydrogen molecules or the structure of the zeolites and is optimized at low temperatures. Zeolites offer a wide range of pore sizes, enabling selectivity in hydrogen adsorption while avoiding diffusion limitations, which is also beneficial for separating hydrogen from other gases in a mixture. The storage capacity of zeolites is highly influenced by pressure and temperature, requiring a balance between maximizing storage capacity and minimizing energy and material costs. Additionally, zeolites exhibit excellent durability, allowing for regenerations after hydrogen desorption without significant degradation, which is crucial for industrial applications aimed at reducing operational costs.

However, zeolites face challenges, including a limited storage density compared to methods like liquefaction or high-pressure compression. Furthermore, the kinetics of adsorption and desorption can be slow, potentially limiting efficiency for certain practical uses, such as refueling hydrogen vehicles. Current research efforts aim to improve the properties of zeolites for hydrogen storage, including developing materials with more suitable pore sizes and enhancing performance under various operational conditions. Surface modifications or combinations with other materials are also being explored to increase their storage capacity and improve adsorption/ desorption rates.

3.3.1.3.3 Metal-Organic Frameworks (MOFs)

MOFs, or metal-organic frameworks (Figure 3.6), are three-dimensional structures composed of metal ions coordinated to organic ligands. These components fall into two main categories: metal centers, which often include ions of zinc, copper, or iron acting as junction points for the ligands, and organic ligands, typically based on carbon structures like bipyridines or carboxylates, forming bridges between the metal centers to create the MOF structure. This configuration gives MOFs very high porosity and a large internal surface area, making them ideal for various applications such as hydrogen storage. They can adsorb significant amounts of gas under low pressures and near-ambient temperatures, making them very promising for sustainable energy applications. In addition to hydrogen storage, these materials are used in CO_2 capture, gas separation, catalysis, and even drug delivery due to their ability to slowly release encapsulated chemicals.

MOFs are primarily fabricated through solvothermal or hydrothermal synthesis, where metal and organic precursors are dissolved in a solvent and heated under pressure, allowing for the orderly coordination of ligands around the metal centers. Other synthesis methods include melt synthesis, electrochemical synthesis, and microwave-assisted synthesis. MOFs have adjustable pore sizes, enabling the selective adsorption of certain gases while excluding others, crucial for applications like separating carbon dioxide from flue gases in CCS technologies. Additionally, the metal active sites in MOFs facilitate various chemical reactions, useful in producing fine chemicals and polymers. In medicine, they allow for controlled and targeted drug release, offering significant potential for delivering treatments such as anticancer drugs. MOFs also serve as sensors for detecting toxins and pollutants due to their variable properties in the presence of specific molecules.

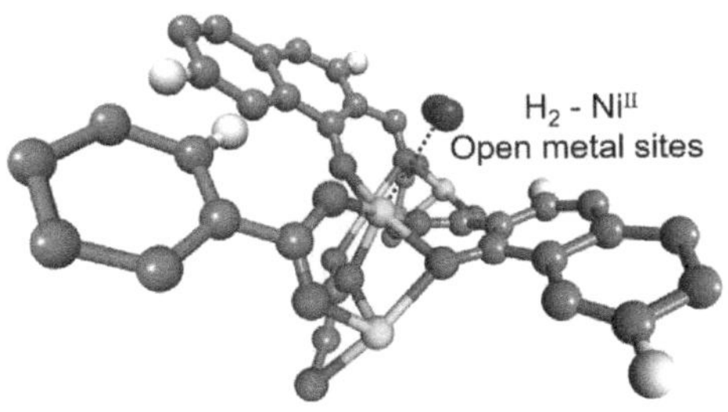

FIGURE 3.6 Hydrogen molecules stored in MOF-74.

(Adapted from ref. [3].)

3.3.1.3.4 Mesoporous Silica

Mesoporous silicas are particularly interesting materials for hydrogen storage due to their specific porous structures. The microstructure of mesoporous silica is characterized by pores with diameters typically ranging between 2 and 50 nanometers. These pores are organized in a regular and uniform manner, creating a large accessible internal surface area. This orderly porous organization is ideal for the physical adsorption of gases, including hydrogen. The performance of mesoporous silicas in hydrogen storage largely depends on their specific surface area and pore size. Hydrogen storage capacity is generally measured by the amount of gas that can be adsorbed at a certain pressure and temperature. Mesoporous silicas can store hydrogen through physisorption, where hydrogen is held on the material's surface by van der Waals forces rather than chemical bonds. Performance can be enhanced by modifying the internal surface through surface treatments or incorporating other elements that can interact more strongly with hydrogen.

The production cost of mesoporous silicas can vary, mainly depending on the sophistication of the synthesis process and the surface modifications required to optimize hydrogen storage. While generally less expensive than other advanced materials like MOFs or CNT, costs can increase with production scale and the need to achieve specific performance levels. The main advantages of mesoporous silicas include a high specific surface area, allowing for a large hydrogen adsorption capacity, uniform, and ordered pores that facilitate predictable and efficient gas adsorption, and chemical and thermal stability, making them suitable for long-term applications in various environmental conditions. However, despite these advantages, mesoporous silicas also present some challenges for hydrogen storage. Their storage density is limited because physisorption does not allow for the very high storage densities comparable to liquid or compressed hydrogen. Additionally, the amount of hydrogen adsorbed decreases significantly at higher temperatures and requires relatively high pressures for effective adsorption. Finally, gas recovery can be complex, as hydrogen desorption may require precise adjustments of temperature or pressure, complicating the design of storage systems.

Mesoporous silicas, with their ordered structure and efficient hydrogen adsorption capacity, offer a promising solution for gas storage. However, improvements in the synthesis and treatment technologies of these materials are necessary to overcome challenges related to storage density and recovery efficiency. Research continues to evolve to optimize these materials for commercial and industrial applications.

3.3.1.3.5 Boron Carbide

Boron carbide is an interesting material for hydrogen storage due to its unique properties. Often designated by its chemical formula B4C, it is an extremely hard and thermally stable ceramic material. Its crystalline structure is composed of boron and carbon units arranged to form a robust three-dimensional network. This structure can be beneficial for creating adsorption sites for hydrogen. In terms of hydrogen storage, boron carbide has shown moderate adsorption capacity. Although its specific surface area is not as high as materials like MOFs or activated carbons, surface treatments or chemical modifications can be used to improve its adsorption performance. Hydrogen storage in boron carbide is primarily achieved through physical interactions, which can limit the amount of hydrogen stored compared to stronger chemical bonds.

The production of boron carbide is relatively expensive due to the high temperatures and specialized equipment required for its synthesis. Costs can also increase depending on the additional treatments needed to optimize the surface for hydrogen storage. Boron carbide offers several notable advantages, including its high hardness and abrasion resistance, making it durable in harsh environments, and its thermal and chemical stability, allowing it to withstand high temperatures and corrosive environments, ideal for demanding applications. Additionally, it exhibits radiation resistance, qualifying it for use in radiation-exposed environments without significant degradation. However, the use of boron carbide for hydrogen storage also presents certain challenges: its hydrogen storage capacity is limited without specific modifications, and it is less effective than other more specialized adsorbent materials. The high production costs, due to the need to produce high-purity boron carbide, as well as the difficulties in handling and shaping due to its extreme hardness, can complicate its integration into certain storage tank designs.

3.3.1.3.6 Carbon Nanotubes

Nanostructured materials, particularly CNTs, represent a promising class for the physical storage of hydrogen through adsorption. Here is a detailed analysis of these materials in terms of microstructure, capacity and performance, costs, advantages, and challenges:

CNTs are single-atomic carbon cylinders with diameters on the nanometer scale and lengths reaching several micrometers. They can be single-walled (SWCNTs) or multiwalled (MWCNTs). The unique structure of CNTs offers an extremely high internal surface area and a network of nanometer-sized pores, facilitating the adsorption and storage of gas molecules. CNTs can store hydrogen at high densities primarily through physical adsorption, where hydrogen is attracted to the internal and external walls of the nanotubes by van der Waals forces. The performance varies based on purity, type of nanotubes, diameter, arrangement, and surface treatment. Studies show that pretreating CNTs to introduce defects or functional groups can significantly increase their hydrogen adsorption capacity.

The production of high-quality CNT is expensive, requiring advanced techniques such as CVD, laser ablation, or electric arc discharge. These methods require specialized equipment and controlled reaction conditions, which increase the overall cost. CNTs offer several significant advantages for hydrogen storage, including a high specific surface area that allows for large storage capacity, structural flexibility that can be adapted to enhance interactions with hydrogen, and chemical and mechanical stability that enables them to withstand harsh environmental conditions and be reused without significant performance loss. However, the use of these nanotubes for hydrogen storage also presents several challenges: their adsorption efficiency decreases at high temperatures, the synthesis methods are complex and costly, handling and storing hydrogen pose safety risks, particularly regarding explosion hazards, and despite their chemical stability, CNT can pose issues in terms of recyclability and environmental impact.

3.3.1.3.7 *Polymer-based Adsorbents*

These are mostly developmental products that currently have very specific and limited applications, primarily due to their cost. The most common is a copolymer of styrene and divinylbenzene. The styrene–divinylbenzene copolymer creates a unique structure with polystyrene chains connected by divinylbenzene bridges. This specific configuration results in significant interchain porosity. During polymerization, the addition of a nonpolymerizable agent creates additional pores, thereby increasing the total porosity of the material. These adsorbents are often shaped into beads with diameters of 1 to 2 mm.

A key property of these adsorbents is their marked hydrophobicity, which makes them particularly effective for the adsorption of volatile organic compounds (VOCs) and potentially interesting for hydrogen storage. The hydrophobicity helps prevent water saturation, allowing for better performance in gas adsorption, including hydrogen. The materials can be used in their initial form or carbonized to form adsorbents similar to activated carbon, which have a larger surface area and improved adsorption capacity. If these polymers are transformed into fibers, they can be woven to make activated carbon fabrics. These fabrics retain the appearance of traditional fabrics while benefiting from the adsorption properties of activated carbon, with fibers about 10 µm in diameter that accelerate hydrogen transfer compared to other adsorbents. Currently, these materials find small-scale applications such as odor capture or as liners for protective clothing against chemical weapons.

An innovative aspect of these fabrics is their electrical conductivity, allowing them to be heated by the Joule effect. This property is exploited to regenerate the adsorbent: the generated heat causes the desorption of hydrogen or other captured gases, making the material reusable. This method of regeneration by temperature elevation represents a significant advancement in hydrogen storage technology R&D. The main challenge remains in the production cost of these specialized polymers and their adaptation to broader applications, along with technical challenges related to maximizing hydrogen storage capacity and regeneration efficiency.

3.3.1.3.8 *Cryoadsorption*

Cryoadsorption represents a promising method for hydrogen storage, conceivable as a complement to liquefaction storage systems. This technique utilizes low temperatures to enhance the efficiency of hydrogen storage. In practice, when liquid hydrogen evaporates, it produces a low-temperature gas that can be immediately captured by cryoadsorption. Interestingly, if the tanks used for this adsorption are designed to withstand pressure, simply warming them to room temperature—for example, by removing thermal insulation—can transform the adsorbed hydrogen into compressed hydrogen. This approach could integrate three storage modes: liquefaction, adsorption, and gas compression.

Advanced research in this field has been conducted at the University of Lorraine, where hydrogen compression from cryoadsorption has been modeled. Furthermore, recent molecular simulations predict that the use of functionalized graphenes, such as graphanes, could double the mass storage capacity of hydrogen. This gain is achieved through a combination of physisorption and covalent bonding of hydrogen to graphene, facilitated by optimal thermodynamic conditions, such as low temperature and high pressure. These conditions are enhanced by functionalizing graphene with hydrogen atoms, promoting efficient molecular adsorption on the structured surface of graphanes.

Additionally, there have been notable developments in the creation of hierarchical nanoporous membranes (HNMs), a class of nanocomposites that combines carbon spheres with graphene oxide. These HNMs exhibit a complex porous structure, combining ultra-micropores and mesopores, which allow for a hydrogen adsorption capacity of up to 3.3% by weight under conditions of 77 K and 1.2 bar. These advancements indicate significant progress in hydrogen storage technologies, offering prospects for improving efficiency and storage capacity.

3.3.1.4 Comparison of Various Adsorbents for Physical Hydrogen Storage

Table 3.4 compares various adsorbents in terms of hydrogen adsorption capacity, production methods, costs, advantages, and challenges. Here is a summary table of these aspects. This table provides an overview of the capabilities and limitations of each type of adsorbent, highlighting their potential in the field of hydrogen storage and the associated technical and economic challenges.

3.3.2 LIQUID HYDROGEN STORAGE

The storage of hydrogen in liquid form is a major technological advancement that meets the needs for increased energy density and logistical efficiency in various sectors. The liquefaction of hydrogen at −253°C allows for a significant reduction in its volume, thus facilitating its transportation and storage in large quantities. This property is particularly advantageous for applications such as refueling hydrogen vehicles, space travel, and certain industrial uses where hydrogen can serve as a raw material or energy source. However, maintaining hydrogen in this liquid state requires advanced storage technologies, which must minimize thermal leakage to prevent hydrogen evaporation. The reservoirs used are typically double-walled containers, often made of stainless steel, with a vacuum between the walls to provide thermal insulation. Inside, heat-reflective materials such as Mylar layers or aluminum are used to further reduce heat losses through radiation. Safety is another major concern, as hydrogen is highly flammable and can react violently in the presence of oxygen. Therefore, rigorous safety measures are essential to prevent fire and explosion risks. This includes sophisticated leak detectors, specially designed ventilation systems, and strict protocols for handling and maintaining the reservoirs. Despite these challenges, the potential of liquid hydrogen as a clean energy carrier is immense. Research continues to evolve in the development of lighter, more efficient, and less-expensive storage materials, as well as energy-efficient liquefaction and regasification methods, which could revolutionize the way energy is stored and used worldwide.

3.3.2.1 The Mechanism of Hydrogen Liquefaction

The storage of hydrogen in liquid form represents a significant technological advancement that addresses the need for increased energy density and logistical efficiency

TABLE 3.4

A Comparison of Different Adsorbents

Type of Adsorbent	Adsorption Capacity (m^2/g)	Production Method	Advantages	Challenges
Activated Carbons	500–1500	Thermal decomposition and chemical or steam activation	High porosity, versatility, availability	Management of by-products, need for regeneration
Zeolites	300–1000	Synthesis from silica and alumina	High selectivity, excellent durability, effectiveness at low temperature	Limited storage density, slow adsorption/desorption kinetics
MOFs (Metal-Organic Frameworks)	1000–5000	Solvothermal, hydrothermal synthesis, or others	Very high porosity, extremely high specific surface area	High cost, moisture sensitivity, production complexity
Mesoporous Silicas	800–1500	Synthesis with structure-directing agents	Chemical and thermal stability, high specific surface area	Limited storage density, production costs for high specifications
Boron Carbide	200–800	High-temperature synthesis	Temperature and abrasion resistance, chemical stability	Very high production cost, difficult handling
Carbon Nanotubes	500–2200	CVD, laser ablation, arc discharge	Extreme specific surface area, mechanical strength, structural flexibility	High production costs, recycling and safety issues
Polymer-Based Adsorbents (e.g., Styrene-Divinylbenzene Copolymer)	300–1200	Synthesis of copolymers, sometimes followed by carbonization	Hydrophobicity, potential for easy regeneration, form adaptability	Still limited development, adaptation costs for large-scale production
Cryoadsorption	Variable	Variable	Effective use of low temperatures, can be coupled with other storage methods	Management of low temperature, need for specific insulation

across various sectors. Liquefying hydrogen at −253°C enables a substantial reduction in volume, thereby facilitating its transportation and storage in large quantities. This characteristic is especially beneficial for applications like refueling hydrogen vehicles, space travel, and specific industrial uses where hydrogen serves as either a raw material or an energy source.

However, maintaining hydrogen in its liquid state necessitates advanced storage technologies that minimize thermal leakage to prevent hydrogen evaporation. Typically, reservoirs used for this purpose are double-walled containers, often constructed from stainless steel, with a vacuum between the walls to provide thermal insulation. Additionally, internally, heat-reflective materials such as Mylar layers or aluminum are employed to further diminish heat losses via radiation. Safety represents another significant concern, given hydrogen's highly flammable nature and its potential for violent reactions in the presence of oxygen. Consequently, stringent safety measures are imperative to mitigate fire and explosion risks. These measures encompass sophisticated leak detection systems, specially engineered ventilation setups, and strict protocols governing the handling and maintenance of reservoirs.

Despite these challenges, the potential of liquid hydrogen as a clean energy carrier is vast. Research endeavors persist in the development of lighter, more efficient, and cost-effective storage materials, along with energy-efficient liquefaction and regasification techniques. These advancements have the potential to revolutionize energy storage and utilization on a global scale.

3.3.2.2 Thermodynamic Models for Liquid Hydrogen Storage

To model the storage of liquid hydrogen, several thermodynamic approaches can be employed, depending on the aspects one wishes to explore (such as storage capacity, evaporation losses, temperature regulation, etc.).

3.3.2.2.1 Model Based on the van der Waals Equation of State

This equation accounts for interactions between molecules and the finite volume of hydrogen molecules. For hydrogen, the van der Waals equation can be expressed based on Equation (3.13):

$$\left(P + a\left(\frac{n^2}{V^2} \right) \right)\left(V - nb \right) = nRT \tag{3.13}$$

where P is the pressure, V is the volume, n is the number of moles of hydrogen, R is the ideal gas constant, T is the temperature, and a and b are constants specific to hydrogen that reflect the attractive forces between molecules and the excluded volume of molecules, respectively.

The van der Waals equation modifies the ideal gas law to account for the nonideal behavior of real gases. The term $a\left(\dfrac{n^2}{V^2} \right)$ corrects for intermolecular attractions, and nb corrects for the volume occupied by gas molecules. This model provides a more accurate description of hydrogen gas behavior, especially under conditions where deviations from ideal gas behavior are significant.

3.3.2.2.2 Model based on the Redlich–Kwong Equation of State

This equation is an improvement over the van der Waals equation and is particularly useful for calculations at high temperatures and low pressures, but can be adjusted for other conditions according to Equation (3.14):

$$P = \frac{RT}{V-b} - \frac{a}{\sqrt{T}V(V+b)} \tag{3.14}$$

where the variables are the same as for the van der Waals equation, a and b are adjusted to better match the properties of liquid hydrogen.

The Redlich–Kwong equation refines the van der Waals equation by introducing a temperature dependency in the attraction parameter aa. This makes the equation more accurate for predicting the behavior of gases over a wider range of temperatures and pressures, providing better results for hydrogen, especially at high temperatures and low pressures.

3.3.2.2.3 Evaporation Loss Model

Evaporation losses (or "boil-off") are critical in the storage of liquid hydrogen. A simple model to calculate the evaporation rate can be based on the energy balance according to Equation (3.15):

$$m.h_{fg} = U.A.\left(T_{amb} - T_{liq}\right) \tag{3.15}$$

where m is the mass evaporation rate, h_{fg} is the enthalpy of vaporization, U is the overall heat transfer coefficient, A is the contact surface area, T_{amb} is the ambient temperature, and T_{liq} is the liquid temperature.

This model calculates the rate of hydrogen evaporation due to heat transfer from the surroundings. The energy transferred to the liquid hydrogen increases its temperature and causes some of it to evaporate. The equation balances the heat gained by the liquid hydrogen with the energy required for phase change (vaporization).

3.3.2.2.4 Overall Thermal Balance

For long-term storage, a thermal balance model can be used to estimate the insulation and cooling requirements according to Equation (3.16):

$$\Delta E_{storage} = Q_{in} - Q_{out} \tag{3.16}$$

where Q_{in} is the ingoing heat, Q_{out} is the outgoing heat, and $\Delta E_{storage}$ is the change in energy in the storage.

This model tracks the energy balance within a hydrogen storage system. By accounting for the heat entering and leaving the system, it helps in determining the insulation and cooling requirements to maintain the desired storage conditions. Proper thermal management ensures the stability and efficiency of hydrogen storage, especially over extended periods.

To conclude, these models can be adapted and combined based on the specific requirements of the storage facility and operating conditions. For more detailed analyses, specialized thermodynamic simulation software can be used to simulate more complex and dynamic scenarios, providing deeper insights into the behavior and management of hydrogen storage systems.

3.3.2.3 Available Technologies for Hydrogen Liquefaction

Liquefaction of hydrogen is an essential process for its handling, storage, and transportation, as it significantly reduces its volume. The main technologies used for hydrogen liquefaction and the challenges associated with each method are presented as follows.

3.3.2.3.1 Claude Cycle

The Claude cycle is a predominant method for hydrogen liquefaction, appreciated for its efficiency and reliability. This complex process involves several key steps to convert gaseous hydrogen into liquid. Initially, hydrogen is compressed to increase its pressure, which also raises its temperature. Following this compression, hydrogen undergoes a series of expansions through turbines, a technique that plays a crucial role in lowering its temperature. These successive expansions are essential to reach the low temperatures required for hydrogen liquefaction.

One of the main advantages of the Claude cycle is its relatively high energy efficiency. This method allows for more efficient liquefaction of hydrogen in terms of energy consumed compared to other processes. However, despite its efficiency advantages, the Claude cycle requires a considerable initial investment. The infrastructure and specialized equipment required to implement this process represent a significant cost. Additionally, the system requires regular maintenance to ensure optimal operation and longevity, leading to ongoing operating costs. Although the Claude cycle is costly both in terms of initial investment and maintenance, its energy efficiency advantages make it an attractive solution for large-scale hydrogen liquefaction. This multistep process, though complex, is crucial for effectively meeting the demand for liquid hydrogen for various industrial applications.

3.3.2.3.2 Joule–Thomson Cycle

The Joule–Thomson cycle is another method used for hydrogen liquefaction, often complementing other techniques. This process focuses on the expansion of compressed hydrogen through a special valve, known as the throttling valve. When hydrogen passes through this valve, it undergoes a pressure drop resulting in rapid cooling of the gas. This temperature reduction is essential for bringing hydrogen to the liquid state as part of this cycle.

One of the major advantages of this cycle is its relative simplicity. Compared to other more complex methods, the Joule–Thomson cycle requires less specialized equipment and can be implemented more easily. This simplicity can reduce initial costs as well as technical requirements for operators. However, the Joule–Thomson cycle has disadvantages, especially regarding efficiency with hydrogen. This gas has unique characteristics that make it less receptive to this type of liquefaction compared to other gases. The main difficulty lies in the fact that hydrogen requires specific conditions, such as extremely low temperatures, to be efficiently liquefied via the Joule–Thomson cycle. Therefore, while this process is useful for its simplicity, it often needs to be combined with other methods to achieve efficient hydrogen liquefaction. The Joule–Thomson cycle, with its operational and setup simplicity, offers a significant advantage for specific applications. Nevertheless, its limitations in terms of efficiency for hydrogen often require it to be employed in tandem with other, more complex processes to achieve optimal results.

3.3.2.3.3 Brayton Cycle

The Brayton cycle and magnetocaloric technologies are two distinct methods used for cooling, including applications involving hydrogen. The Brayton cycle uses a working gas, typically helium, which is compressed and then expanded to produce an indirect cooling effect on hydrogen. This method is particularly effective for large-scale installations, although it is complex and costly to implement. On the other hand, magnetocaloric technology relies on materials that change temperature in response to a magnetic field; they heat up when the field is applied and cool down when it is removed. This approach offers the potential for highly efficient and energy-saving cooling cycles. However, it is still in the early stages of development and research, highlighting challenges related to its large-scale commercialization. Each technology has its specific advantages that make it suitable for certain applications, although costs and technical complexity remain significant considerations.

The liquefaction of hydrogen, while crucial for its storage and transportation, presents several significant challenges that require ongoing technological improvements. This process is particularly energy-intensive, and increasing the energy efficiency of liquefaction equipment remains a major obstacle. The initial installation of these equipment also involves high investment costs. Additionally, these systems require regular maintenance to prevent failures, which can be costly and technically complex. From a safety standpoint, hydrogen, being highly flammable, must be handled with extreme precautions, especially in its liquid form at very low temperatures. Finally, another notable issue is hydrogen loss due to evaporation during storage and transportation, resulting in material and energy losses. Despite these challenges, improving liquefaction technologies is essential to make hydrogen storage and transportation more economical and efficient, especially for large-scale energy applications.

3.3.2.4 Different Types of Tanks Used for Cryogenic Storage

There are several types of tanks used for the cryogenic storage of hydrogen, each with its own advantages and disadvantages.

3.3.2.4.1 Double-walled Stainless Steel Tanks

Double-walled stainless steel tanks for liquid hydrogen storage are designed to provide optimal thermal insulation and increased safety. Their structure consists of two layers of stainless steel, one internal and one external, between which a vacuum is maintained. This vacuum plays a crucial role by acting as a thermal insulator, preventing heat from the external environment from reaching the liquid hydrogen maintained at a temperature of approximately $-253°C$. This feature is essential for minimizing hydrogen losses through evaporation and for maintaining the energy efficiency of the storage system.

The typical shape of these tanks is cylindrical, a configuration that optimizes both structural strength under pressure and efficient use of space. The ends of the tanks are often hemispherical or torispherical, which helps evenly distribute internal pressure and reduce stress points, thereby increasing durability and safety. This shape also allows for easier installation and integration into various configurations, whether for stationary installations or for mobile applications, such as hydrogen vehicles.

The manufacturing of double-walled stainless steel tanks generally begins with the selection of high-quality stainless steel capable of withstanding extremely low temperatures and corrosion. The process includes forming the steel layers into the desired shape, often through stamping or welding formed steel sheets. Once the two shells are formed, they are assembled to create a space between them, which is then evacuated to create the necessary vacuum. This vacuum is crucial as it eliminates thermal conduction and convection between the interior and exterior of the tank. Finally, monitoring systems and safety valves are installed to maintain the tank's performance and safety over time.

One of the main advantages of double-walled stainless steel tanks is their robustness. Stainless steel is known for its fracture resistance and durability over time, which is essential in storage conditions where safety is paramount. Additionally, these tanks provide good thermal insulation due to their double-walled design, essential for maintaining liquid hydrogen at $-253°C$. Economically, although their manufacturing is relatively expensive, their longevity and reliability can justify this initial investment, making operational costs more manageable in the long run.

However, double-walled stainless steel tanks also have several significant disadvantages. Their high weight is one of the main issues, as it can limit their use in applications where weight is a critical constraint, such as in vehicles or certain aerospace infrastructures. Their large size can also be a challenge during installation or integration into existing spaces. Furthermore, although stainless steel is generally corrosion-resistant, under specific conditions, especially in the presence of chlorides and acidic environments, it can undergo corrosion. This susceptibility requires regular maintenance to avoid failures that could have serious consequences, thereby increasing maintenance costs and potentially total ownership costs.

In conclusion, while double-walled stainless steel tanks are a proven solution for liquid hydrogen storage, they are not without faults. The choice of these tanks should therefore be carefully evaluated based on the specific requirements of each application, considering both their advantages in terms of robustness and insulation, and their disadvantages related to weight, size, susceptibility to corrosion, and maintenance requirements.

3.3.2.4.2 Double-walled Aluminum Tanks

Double-walled aluminum tanks for liquid hydrogen storage represent an interesting technological solution, combining lightness and corrosion resistance with relatively moderate production costs. These characteristics make them a viable option, particularly in industries where weight is a critical factor, such as in the automotive and aerospace sectors.

The standard shape of double-walled aluminum tanks is generally cylindrical, with rounded ends for better distribution of internal pressure. This cylindrical configuration maximizes space utilization while providing excellent structural strength under pressure. The double-walled construction plays a crucial role in the tank's thermal insulation. Between the two aluminum walls, a vacuum space is maintained to minimize heat transfer, crucial for keeping hydrogen in a liquid state at extremely low temperatures ($-253°C$).

The manufacturing of double-walled tanks begins with the formation of two aluminum shells. Aluminum is chosen for its ease of handling and forming, often through extrusion or hot forming methods. After the formation of the shells, they are assembled to form a double wall with an intermediate space. This space is then evacuated, and reflective barriers or insulating materials may be added to enhance insulation. Joints and connections are carefully welded and tested to ensure the tank's tightness and resistance under the extreme temperature and pressure conditions it will be exposed to.

Double-walled aluminum tanks are valued for their lightweight, reducing the overall weight of systems that incorporate them, thereby increasing the energy efficiency of vehicles or hydrogen transport systems. Aluminum is also highly corrosion-resistant, extending the life of the tanks and reducing maintenance needs related to corrosion. The manufacturing cost of these tanks, while higher than that of tanks made from composite materials, remains reasonable compared to more expensive options such as stainless steel.

However, aluminum tanks are not without drawbacks. They offer less efficient thermal insulation than stainless steel tanks, which may require more complex or frequent cooling systems to maintain hydrogen in a liquid state. Additionally, although aluminum is corrosion-resistant, it may require additional surface treatments, especially in aggressive environments or for specific applications. Finally, aluminum can be susceptible to stress cracking, especially when exposed to repeated temperature cycles or dynamic loads, which can compromise the tank's structural integrity over time.

Given the above-mentioned explanations, while double-walled aluminum tanks offer significant advantages in terms of weight and corrosion resistance, they require special attention during design and maintenance to maximize their performance and durability in liquid hydrogen storage applications. These tanks continue to be developed and improved to meet the increasingly strict requirements of modern energy applications.

3.3.2.4.3 *Vacuum Insulation Tanks (VLI) in Steel or Aluminum*

Vacuum Insulation Tanks (VLI) in steel or aluminum represent advanced technology for liquid hydrogen storage, offering significant benefits for applications requiring high reliability and performance. These tanks use a vacuum insulation principle to maintain hydrogen at extremely low temperatures, thereby minimizing hydrogen losses through evaporation. The design of these tanks typically includes two walls between which the vacuum is created, providing an effective thermal barrier. The manufacturing of these tanks requires high technical expertize. Steel or aluminum is often used for the construction of the internal and external walls due to their corrosion resistance and ability to withstand low temperatures. After the walls are formed, a vacuum is carefully created between them, and special insulation materials, such as reflective multilayer panels, are sometimes added to enhance insulation efficiency. This multi-step construction contributes to the complexity and high cost of VLI tank manufacturing.

The advantages of these tanks are notable. They offer excellent thermal insulation, which is crucial for reducing hydrogen losses through evaporation, a common phenomenon when storing hydrogen at cryogenic temperatures. Furthermore, while robust, these tanks are designed to be relatively lightweight and compact, facilitating their integration into various transportation or energy storage systems.

However, VLI tanks are not without disadvantages. Their initial cost is generally higher compared to other types of tanks, mainly due to the complexity of their construction and the special materials required for vacuum insulation. Additionally, their structure can be relatively fragile, requiring extra precautions during handling and transportation.

VLI in steel or aluminum are highly effective solutions for liquid hydrogen storage, suitable for applications where safety and efficiency are paramount. Despite their cost and complexity, their ability to minimize hydrogen losses and superior thermal performance justify the initial investment for many industrial and research projects.

3.3.2.4.4 *Carbon Fiber Reinforced Composite Tanks*

Carbon fiber-reinforced composite tanks represent cutting-edge solutions for liquid hydrogen storage, particularly valued in sectors where lightness and performance are critical, such as automotive or aerospace. These tanks are distinguished by their composite structure, which incorporates carbon fibers integrated into a polymer matrix. This blend provides the tank with exceptional strength while reducing its weight, a considerable asset for both mobile and stationary applications.

The manufacturing of these tanks is a complex technical process that begins with the strategic placement of carbon fibers on a mold. These fibers are then impregnated with a polymer resin that hardens to form the structural matrix of the tank. This process, known as resin transfer molding or filament winding, allows for the creation of tanks that are not only lightweight but also capable of withstanding the high pressures and low temperatures required for liquid hydrogen storage.

The advantages of these tanks are numerous. They are extremely lightweight, reducing the overall vehicle load and thereby increasing energy efficiency. Additionally, the corrosion resistance of carbon fibers ensures increased longevity even in hostile environments. These tanks also offer very good thermal insulation, crucial for minimizing hydrogen losses through evaporation.

However, these composite tanks are not without disadvantages. Their high cost can be a major obstacle to wider adoption. They are also susceptible to damage from impacts, which can compromise their structural integrity and require additional protective coatings to mitigate this risk. Moreover, composite materials may degrade over time, especially due to exposure to ultraviolet environments or extreme temperatures, which can reduce their effectiveness and necessitate regular monitoring and maintenance.

In conclusion, carbon fiber reinforced composite tanks represent advanced technology for liquid hydrogen storage, offering an optimal combination of lightness, strength, and thermal insulation. Despite their high costs and sensitivity to impacts, their superior performance makes them indispensable in many applications where efficiency and durability are paramount.

3.3.2.4.5 *Solid Material Adsorption Tanks*

Solid material adsorption tanks represent an innovative approach to hydrogen storage, leveraging the properties of adsorbent materials to capture and store hydrogen in gaseous rather than liquid form. This technology utilizes porous materials such as activated carbons, molecular sieves, or MOFs, which have the ability to fix large quantities of hydrogen on their internal surface at moderate pressures.

The manufacturing of these tanks begins with the selection and preparation of the adsorbent material, which is then conditioned and placed inside a robust tank designed to withstand the pressure and temperature variations inherent in hydrogen adsorption and desorption. The material is often configured to maximize the available internal surface area, which is crucial for increasing the tank's hydrogen storage capacity.

One of the main advantages of these tanks is their ability to store a large amount of hydrogen without the need to maintain cryogenic temperatures, thereby eliminating heat losses and the need for external cooling systems. This makes them particularly attractive for applications where simplicity and reduction of operational costs are essential.

However, adsorption tanks also have some drawbacks. Their volumetric hydrogen storage density is generally lower than that of liquid or compressed gas tanks, which can limit their usefulness in applications where space is a critical factor. Additionally, loading and unloading hydrogen require specific temperature and pressure conditions, which can complicate their integration into certain energy systems. Finally, the adsorption and desorption kinetics can be relatively slow, restricting the speed at which hydrogen can be extracted for immediate use.

In summary, solid material adsorption tanks offer a promising solution for hydrogen storage, with significant advantages in terms of simplicity and operational costs. However, their limitations in terms of storage density and loading/unloading dynamics must be taken into account when evaluating their adaptability to specific applications.

Each type of tank has its own characteristics in terms of performance, cost, weight, and manufacturing complexity. The choice of tank often depends on the specific requirements of the application, such as mobility, storage capacity, safety, and available budget.

3.3.2.5 The Liquid Hydrogen Industry

Liquid hydrogen is used in several industrial sectors, where its unique properties such as being lightweight, highly energetic, and completely clean during combustion (only releasing water) are particularly valued. Aerospace missions to outer planets and other distant destinations often use launchers equipped with engines running on liquid hydrogen to benefit from its high energy efficiency, which is crucial for reaching the high speeds needed to escape Earth's gravitational pull. Liquid hydrogen is also used as fuel for rockets, providing a very efficient and powerful source of energy essential for satellite launches and manned space missions. Additionally, in the chemical industry, hydrogen serves as a reactant in the production of various chemicals, including NH_3 and methanol, which are precursors to many other chemicals and fertilizers. Hydrogen is also essential in the oil refining process, where it is used to hydrotreat petroleum fractions, helping to remove sulfur and convert hydrocarbons into lighter products such as gasoline and diesel.

In the energy sector, liquid hydrogen serves as an energy carrier in energy storage and transportation, particularly for storing surplus energy produced by intermittent renewable sources such as wind or solar power. It can be converted back into electricity via fuel cells during periods of high demand and is also being considered for grid balancing and as a decarbonization method for energy-intensive industries. While less common, the transportation sector also benefits from the storage of hydrogen in liquid form, particularly in specialized sectors such as heavy transport and aerospace vehicles, contributing to the reduction of emissions associated with traditional fuels like CO_2, NO_x, and SO_x. Additionally, some experimental ships use hydrogen as fuel, emitting primarily water and thus reducing emissions associated with traditional fuels.

R&D laboratories utilize liquid hydrogen for various research applications, including low-temperature physics and studies on superconductors, as well as for developing new materials such as metal hydrides for hydrogen storage or materials for fuel cells. In electronics, hydrogen is used in the production of semiconductors and other electronic components as a purging and cleaning gas to remove contaminants. It is also used in crystal growth processes for semiconductors, acting as a carrier gas to transport chemical precursors to the substrate surface where the crystal forms. Furthermore, in the food sector, liquid hydrogen can be used as a refrigerant for preserving and transporting sensitive food products, and it is also used in the process of hydrogenating vegetable oils, transforming liquid oils into solid fats at room temperature, such as margarine and other types of hardened vegetable fats.

In metallurgy, liquid hydrogen is employed as a protective atmosphere for the reduction and purification of certain metals, as well as in the brazing process under a controlled atmosphere to prevent oxidation of metals during brazing, a method of joining metal parts by melting a filler metal between them. Each application benefits from hydrogen's ability to release a large amount of energy or provide specific conditions suitable for specific processes. The transition to a greener economy could see an expansion of these uses, especially in the energy and transportation sectors.

3.3.2.6 Recent Advances and Practical Applications in Cryogenic Storage

Research on cryogenic hydrogen storage has seen several significant advances recently, which have the potential to improve both the efficiency and reliability of hydrogen storage systems. Among these advancements, one of the most promising involves the development of new materials for thermal insulation. Work is underway to harness the exceptional properties of aerogels and nanomaterials, which could significantly reduce hydrogen losses through evaporation while enhancing the efficiency of cryogenic tanks. Simultaneously, progress has also been made in containment technologies. Carbon fiber-reinforced composite materials, for example, are being studied for their ability to enhance the safety of cryogenic tanks. These materials promise to reduce the risks of leakage and rupture, which are crucial for maintaining safe and reliable storage. Additionally, the industry is exploring innovative thermal management systems, such as heat recovery systems and advanced heat exchangers. These technologies aim to optimize the use of thermal energy and minimize hydrogen losses, thereby strengthening the overall efficiency of cryogenic storage.

Detection and monitoring technologies have also benefited from notable advances. The development of more sensitive leak detection sensors and improved real-time monitoring systems plays a crucial role in incident prevention, enabling rapid and effective detection and intervention in the event of hydrogen leakage. Finally, the adoption of advanced modeling and simulation techniques enriches our understanding of the thermal and dynamic processes associated with cryogenic storage. These simulation tools are essential for optimizing the design of storage systems and predicting their behavior under various operational conditions.

By integrating these innovations into the design and operation of hydrogen storage systems, we can not only increase their efficiency but also their reliability and safety. These improvements are essential for promoting the wider use of hydrogen as a clean and renewable energy source, paving the way for more sustainable and environmentally friendly energy applications.

3.3.2.7 Environmental and Economic Considerations of Cryogenic Hydrogen Storage

Cryogenic hydrogen storage is a key technology for energy transition, but it raises several environmental and economic considerations. On the environmental front, GHG emissions from hydrogen production are a major concern, especially when hydrogen is produced from fossil fuels. These emissions can be mitigated by using renewable energy sources, such as water electrolysis, and by CCS technologies. Additionally, the processes of hydrogen liquefaction require considerable energy, thus contributing to emissions if this energy is not obtained from renewable sources. There is also an impact on local ecosystems, particularly in terms of land consumption and natural resource use for storage infrastructure. To integrate cryogenic hydrogen storage into existing infrastructures, especially for transportation and industrial applications, careful planning and adaptation of facilities are necessary. This integration must assess the specific needs of different sectors, adapt existing infrastructure for compatibility with cryogenic hydrogen, and develop new dedicated infrastructures. Integration into existing distribution networks and the promotion of collaboration between industry stakeholders and governments are essential for an effective transition.

On the economic front, although the initial investment costs and operating costs of cryogenic hydrogen storage are high, they are offset by potential economic benefits. Hydrogen offers an alternative to fossil fuels, thereby reducing costs associated with GHG emissions and energy price fluctuations. Additionally, the development of this industry can stimulate job creation and technological innovation, contributing to long-term economic benefits.

3.3.3 Compressed Hydrogen Storage

The storage of hydrogen in the form of compressed gas is a crucial technique for managing and utilizing hydrogen as a clean energy source. As the world seeks to reduce its reliance on fossil fuels, hydrogen is emerging as a promising alternative, partly due to its ability to be efficiently stored and transported in compressed form. Compressed hydrogen gas is typically stored at high pressure, often between 350 and 700 bars, in reservoirs specially designed to withstand such stresses and minimize the risks of leakage or explosion. This compression allows for a greater quantity of hydrogen to be stored in a reduced volume, making it a preferred method for applications requiring high energy density, such as transportation or energy distribution. However, compressing hydrogen poses several technical and safety challenges. The materials used for the reservoirs must not only withstand high pressures but also be impermeable to hydrogen to prevent leakage. Additionally, the compression processes require energy, raising questions about the overall energy efficiency of compressed hydrogen storage.

Despite these challenges, compressed hydrogen gas storage plays a vital role in integrating hydrogen into our energy matrix. Whether powering fuel cell vehicles, storing energy for electrical grids, or serving as an energy carrier in industrial processes, compressed hydrogen offers a flexible and adaptable solution for various energy applications. However, the success of its deployment will depend on optimizing the compression technology and continuously improving safety standards and leak detection methods, thereby ensuring the safe and efficient use of this valuable resource.

Compressing hydrogen gas involves complex considerations to ensure its safety and effectiveness for various applications. From understanding the thermodynamics of compression processes to selecting suitable materials for high-pressure storage containers, scientific research plays a vital role in advancing hydrogen compression technologies. Additionally, studies focused on optimizing compression methods and enhancing leak detection systems contribute to improving the overall reliability and efficiency of compressed hydrogen storage systems. Through rigorous scientific inquiry and technological innovation, the challenges associated with compressed hydrogen storage can be addressed, paving the way for its widespread adoption in clean energy solutions.

3.3.3.1 Methods and Technologies for Hydrogen Compression for Storage and Transport

3.3.3.1.1 *Piston Compressors*

Piston compressors are widely used for hydrogen compression, mainly due to their ability to achieve extremely high pressures. These compressors operate on a principle similar to that of ICEs but in reverse, using mechanical energy to move one or more pistons within a closed cylinder. The compression process occurs in several stages: during the intake phase, the piston moves back, creating a vacuum that draws hydrogen into the cylinder through an intake valve. During compression, the piston moves up, compressing the hydrogen and reducing its volume, which increases its pressure. Finally, during the discharge phase, the compressed hydrogen is expelled from the cylinder through an output valve to the storage or distribution system.

For applications requiring even higher pressures, piston compressors can be configured in multiple stages, where the hydrogen compressed in the first cylinder is transferred to a second cylinder for further compression. Each stage of the compression cycle gradually increases the gas pressure. Between stages, hydrogen can be cooled to improve the efficiency and safety of the process. These compressors offer significant advantages, including their ability to produce high pressures, which are essential for many industrial applications and high-pressure hydrogen storage. They are also known for their robustness and capability to operate continuously for long periods with proper maintenance. However, they present challenges such as mechanical wear of pistons, cylinders, and valves, requiring regular maintenance to maintain performance and safety. Additionally, managing lubrication is crucial to avoid hydrogen contamination, particularly in applications where gas purity is paramount. Piston compressors also generate significant levels of noise and vibrations, which may require mitigation measures in some contexts. Although piston compressors are a proven solution for hydrogen compression, suitable for a wide range of industrial applications due to their ability to generate high pressures and flexible configuration, their efficiency and reliability heavily depend on regular maintenance and careful management of associated technical challenges.

3.3.3.1.2 Diaphragm Compressors

Diaphragm compressors are specially designed to compress hydrogen without direct contact with moving mechanical parts, ensuring the high purity of the compressed gas. This feature is particularly valuable in industries where gas contamination must be avoided, such as pharmaceutical, food, or advanced technology applications. The operation of these compressors relies on a flexible diaphragm, often made from rubber or special polymers, which acts as a barrier between hydrogen and the compression mechanism. The operation cycle involves several key steps: during the intake phase, a mechanical movement, usually produced by a lever or an eccentric, pulls the diaphragm back, increasing the volume of the compression chamber and allowing hydrogen to flow in through an intake valve. Next, in the compression phase, the diaphragm is pushed forward, reducing the chamber volume and compressing the hydrogen. Finally, the hydrogen compressed to the desired pressure is directed to a storage or distribution system through an output valve.

The advantages of these compressors include the purity of the hydrogen, as the gas does not come into contact with oils or greases, thus remaining pure, which is crucial for applications requiring high purity standards. They also offer a reduction in leakage risks due to the effective sealing of the diaphragm, minimizing hydrogen leakage, which is important for both safety and operational efficiency. Additionally, these compressors are designed to resist corrosion and wear caused by hydrogen, extending their lifespan, and reducing maintenance needs.

3.3.3.1.3 Centrifugal Compressors

Centrifugal compressors are ideal for handling large volumes of hydrogen, frequently used in industrial sectors requiring high gas flow rates at moderately high pressures. Their uniqueness lies in their dynamic compression method, which uses a bladed wheel to accelerate and then decelerate the hydrogen, allowing for continuous and efficient compression. The operation process of these compressors involves several key stages. Initially, hydrogen is drawn in and directed to the center of the bladed wheel, or impeller. Due to the rapid rotation of this wheel, hydrogen is radially accelerated outward, increasing its kinetic energy. Subsequently, hydrogen passes through a diffusion chamber where its speed gradually decreases, converting kinetic energy into pressure energy, thus raising the gas pressure. Finally, the compressed gas is discharged at the desired pressure, ready for storage or distribution.

These compressors offer several significant advantages. They are highly efficient at handling large volumes of gas, making the process more economical on a large scale. Additionally, unlike piston compressors, they have fewer moving parts, reducing maintenance needs and increasing reliability. Their ability to operate continuously is also a major asset for industrial applications requiring a constant supply of compressed hydrogen.

In terms of applications, these compressors are particularly used in the chemical and petrochemical industries for processes requiring large quantities of pressurized hydrogen, such as NH_3 synthesis or natural gas desulfurization. They are also employed in refineries for hydrogenating oils and fats or producing cleaner fuels, and in hydrogen power plants to manage the significant gas flows needed.

However, these systems face certain challenges, particularly concerning pressure limitations. While excellent for moderate pressures, they are not optimal for achieving very high pressures compared to piston compressors. The initial costs also pose a challenge, as installing a centrifugal compressor can be expensive due to its size and technological complexity. In conclusion, centrifugal compressors offer a robust and efficient solution for handling large volumes of hydrogen, with continuous operation capability and reduced maintenance, making them a preferred choice for many industrial applications, even though they are more suited for moderate pressures and require a significant initial investment.

3.3.3.1.4 *Screw Compressors (Rotary lobe)*

Screw compressors, also known as rotary lobe compressors, play a crucial role in hydrogen compression, offering an efficient solution for handling large volumes of gas at intermediate pressures. These compressors use two helical rotors with lobes and grooves that rotate in opposite directions within a tightly sealed casing. The compression process occurs in three key phases: intake, where hydrogen enters the compressor and is trapped in the spaces between the rotor lobes; compression, during which the rotor lobes rotate and gradually reduce the volume of the spaces containing the hydrogen, thus increasing its pressure; and discharge, where the compressed hydrogen is pushed toward the outlet by the continuous movement of the rotors, directed toward storage systems or for direct use.

These compressors offer multiple advantages. They ensure continuous and uniform compression, reducing pulsations and vibrations compared to other types of compressors, such as piston compressors. Their energy efficiency is also high, making them very effective for applications requiring the compression of large quantities of gas. Moreover, screw compressors are known for their robustness and durability, with fewer moving parts prone to rapid wear, which minimizes the need for frequent maintenance.

These compressors are used in various sectors. In the chemical and petrochemical industries, they are ideal for processes requiring constant and large flows of hydrogen, such as chemical synthesis reactions or fat hydrogenation. They are also used in hydrogen refueling stations to compress hydrogen before it is stored in distribution tanks, as well as in energy storage systems, where they help manage hydrogen used as an energy storage medium from renewable sources. However, these compressors present certain challenges, notably the investment and operating costs, which, while economically advantageous in the long term, can be significant initially. Additionally, despite their robustness, they require regular monitoring to ensure the integrity and operational efficiency of the rotors and seals. In summary, screw compressors for compressed hydrogen offer an advantageous solution for applications requiring high and continuous hydrogen flows, with their ability to operate at intermediate pressures, high efficiency, and reliability making them a preferred choice for many industrial and energy processes.

3.3.3.1.5 Scroll Compressors

Scroll compressors, also known as spiral compressors, stand out due to their efficient compression technology and unique design, making them particularly suitable for compressed hydrogen in various applications. These compressors consist of two interlocking spirals, one fixed and one orbiting, which create compression pockets between them without touching. The operation of these compressors involves several essential phases: hydrogen is first admitted through an opening at the center of the fixed spiral. As the orbiting spiral moves around the fixed one, the captured gas pockets move radially outward and gradually decrease in volume, increasing the pressure of the hydrogen. The compressed hydrogen is then expelled to the storage or utilization system.

These compressors offer several significant advantages. Their operation is notably quiet and smooth compared to piston compressors, generating less noise and vibration due to the absence of abrupt movements and banging parts. Their high energy efficiency results from minimizing internal leaks and maintaining almost continuous compression, thus optimizing energy consumption. Additionally, with fewer moving parts and components prone to wear, scroll compressors require less maintenance, reducing long-term operational costs.

Scroll compressors find applications in various fields. They are ideal for supplying hydrogen to fuel cells, essential for electric vehicles or energy storage units, and are used in gas processing industries where precise pressure control and hydrogen purity are critical, such as semiconductor manufacturing or research laboratories. Though less common, they can also be adapted for HVAC and refrigeration systems requiring gas compression. However, these compressors have limitations, particularly their performance under extreme pressures, which are not ideal for achieving very high levels compared to other types of compressors like piston or screw compressors. Additionally, the initial cost may be higher due to their specific design and the precision required in manufacturing. In conclusion, scroll compressors for compressed hydrogen offer a high-performing and economically advantageous solution with quiet operation, good energy efficiency, and reduced maintenance, gaining popularity as new hydrogen applications develop across various industrial and energy sectors.

3.3.3.1.6 *Liquid Ring Compressors*

Liquid ring compressors for compressed hydrogen are a niche solution particularly effective in contexts where gas purity is paramount and contamination must be avoided. This type of compressor uses a liquid, often water, to provide both sealing and compression of the gas. The operation of these compressors relies on an ingenious mechanism: a circular body or chamber houses an eccentric rotor. When the rotor moves, the liquid is centrifugally forced against the chamber walls, forming a liquid ring. Hydrogen is introduced into this chamber, and the rotor, equipped with blades, moves the gas through the liquid ring. As the rotor turns, the blades reduce the volume of the captured hydrogen pockets, compressing the gas. The compressed gas is then extracted, passing through the liquid, which seals and purifies the gas from any particles or contamination.

This system offers several notable advantages. It is particularly effective in preventing hydrogen contamination, avoiding direct contact with oils or greases, thus reducing the risk of cross-contamination. These compressors are also very robust and suited to harsh industrial conditions where gases may be corrosive. Additionally, although the liquid ring must be frequently monitored, the mechanical simplicity of the compressors facilitates general maintenance operations. These compressors are well-suited for several critical applications. They are perfect for the chemical, pharmaceutical, and electronics industries where hydrogen purity is essential for product quality. They are also used in research centers where sensitive experiments require extremely pure gas supplies, as well as in industrial gas processing and recycling to remove impurities before release or reuse.

However, managing the liquid ring requires frequent adjustments to maintain compression efficiency and avoid breakdowns, and these compressors may not be suitable for achieving the very high pressures required in some applications. In summary, liquid ring compressors for compressed hydrogen represent an effective and reliable solution for applications where high gas purity is crucial, offering excellent reliability, efficiency, and adaptability to harsh conditions while avoiding gas contamination.

3.3.3.1.7 *Electrochemical Compression*

Electrochemical compression for compressed hydrogen is an innovative and potentially revolutionary approach in gas compression technology. Although it is not yet widely deployed commercially, this method offers a promising alternative to traditional mechanical compression techniques by utilizing electrochemical reactions to increase hydrogen pressure. This technology operates through several key steps: first, hydrogen is introduced into an electrochemical cell where it is ionized at an electrode (anode), dissociating hydrogen molecules (H_2) into protons (H^+ ions) and electrons. The protons are then transported across a PEM while the electrons travel via an external circuit to another electrode (cathode). This separation generates an electron flow, creating an electric current. At the cathode, the protons and electrons recombine to form hydrogen molecules under increased pressure due to the electrochemical pressure differential generated by the cell.

The advantages of this method are numerous. In terms of energy efficiency, it often surpasses mechanical methods by directly using electrical energy to move and recombine hydrogen ions, minimizing thermal and mechanical losses. The absence of moving parts reduces wear and maintenance needs. Additionally, this technology allows for precise control of the output pressure, essential in applications requiring strict pressure regulation. Potential applications include the compression of hydrogen for storage from renewable energy sources, helping integrate hydrogen as an energy vector in electrical grids. It can also be used to compress hydrogen for fuel cell vehicle refueling stations, offering a more compact and energy-efficient solution. Special environments such as space or underwater, where traditional compressors are less suitable, also represent ideal applications due to the lack of moving parts and the ability to operate in confined spaces.

However, this technology is still in the developmental phase and requires improvements to optimize its efficiency, durability, and scalability. The high initial costs for development and implementation also limit its immediate adoption. In conclusion, electrochemical compression for compressed hydrogen holds considerable potential to revolutionize how hydrogen is compressed and stored, offering significant advantages in terms of efficiency, maintenance, and pressure control. It could play a key role in the future of hydrogen management, especially in the transition toward cleaner and more sustainable energy sources.

Each compression method has its advantages and disadvantages in terms of cost, energy efficiency, compression capacity, maintenance, and hydrogen purity. The choice of method often depends on the specific application requirements, such as the amount of hydrogen to be compressed, the desired pressure, and the context of use (industrial, transportation, etc.). Hydrogen compressors typically operate by reducing the gas volume to increase its pressure, enabling high energy density storage or transport in pressurized pipelines. Each type of compressor offers specific benefits tailored to diverse needs and applications. To provide a detailed comparison of the different methods for compressing hydrogen based on the mentioned criteria, Table 3.5 presents a comparison of the different hydrogen compression methods in terms of volumetric flow rate, output pressure, energy efficiency, leakage rate, stability and lifespan, and total cost. Note that exact figures may vary depending on the manufacturer and model specifications. The values in the table are typical estimates to give a comparative idea of the performance of each type of compressor.

3.3.3.2 Different Types of Tanks Used for Compressed Hydrogen Storage

To store compressed hydrogen, different types of tanks are used, each designed to meet specific safety standards and to optimize storage capacity and durability. The choice of tank type depends on the specific application, weight requirements, cost, and expected lifespan. Each type of tank has its own advantages and disadvantages, and the choice often depends on the trade-off between cost, weight, and performance required by the specific application, such as hydrogen vehicles, stationary storage, or mobile industrial applications.

3.3.3.2.1 *Hydrogen Storage Tank Type I*

Type I tanks for storing compressed hydrogen, primarily made of steel or aluminum, are distinguished by their simple design and moderate cost, making them attractive for various uses. Their fully metallic structure ensures robustness and resistance, which is crucial for the safety of storage under high pressures. However, the

TABLE 3.5

Comparison of Different Compressors for Compressed Hydrogen Storage[a]

Compressor Type	Volumetric Flow Rate (Nm^3/h)	Output Pressure (bar)	Energy Efficiency (%)	Leakage Rate (%)	Stability and Lifespan (years)	Total Cost (approximate, USD)
Piston Compressors	10–1000	Up to 1000	60–70	<1	10–20	Medium to High
Diaphragm Compressors	5–500	Up to 900	70–80	<0.5	15–25	High
Centrifugal Compressors	100–10000	20–200	75–85	0.5–1.0	20–30	High
Screw Compressors	50–5000	10–100	65–75	1–2	10–20	Medium
Scroll Compressors	1–200	Up to 50	70–80	<1	10–15	Medium
Liquid Ring Compressors	20–2000	1–20	50–60	1–3	10–20	Low to Medium
Electrochemical Compression	<1–100	Up to 1000	80–90	<0.1	5–15	Very High

[a] Abbreviations:

- **Volumetric Flow Rate**: Measures the amount of hydrogen the compressor can process per unit of time. The higher the figure, the more gas the compressor can handle quickly.
- **Output Pressure**: The maximum pressure the compressor can push the hydrogen to. Piston and diaphragm compressors are particularly effective at achieving high pressures.
- **Energy Efficiency**: Indicates how much energy is used effectively. Centrifugal and electrochemical compressors are often the most efficient.
- **Leakage Rate**: Important for safety and efficiency; low values are preferable.
- **Stability and Lifespan**: Evaluates the long-term durability and reliability of the compressor.
- **Total Cost**: Includes the cost of purchase, installation, and maintenance. Varies widely depending on technology and capacity.

main drawback of these tanks lies in their significant weight, a notable handicap for mobile applications, such as hydrogen vehicles, where reducing weight is essential for improving energy efficiency. Despite this handicap, Type I tanks are valuable in several industrial sectors that prioritize robustness, such as fixed hydrogen refueling stations and various manufacturing processes in the chemical and metallurgical industries where hydrogen serves as a reactant or reducing agent. They are also integrated into power generation systems and used as shielding gas in welding, benefiting from their strength despite the compromise on weight.

These tanks remain relevant in industry, especially in chemistry and petrochemistry for storing hydrogen needed for refining and chemical production, as well as in energy production and storage. For the transport and distribution of industrial gases, their durability and easy manufacturing in large sizes are appreciated. Laboratories and research centers also employ them for small-scale experiments, where more sophisticated tanks are not required. Although more advanced tank technologies like Types III and IV are increasingly favored for applications requiring significant weight reduction, Type I tanks remain a reliable and economical solution in contexts where lightness is not a priority.

3.3.3.2.2 Hydrogen Storage Tank Type II

Type II tanks for compressed hydrogen are essential for storing hydrogen under high pressure, typically between 350 and 700 bar. They are mainly constructed of aluminum or steel and reinforced with composite materials such as carbon or glass fibers in the midsection. This hybrid structure retains the strength of metal while benefiting from the lightweight nature of composites, although the reinforcement is less extensive than in Types III and IV tanks, which are primarily or entirely made of composites. One of the main advantages of Type II tanks is their cost-effectiveness. They offer comparable strength and safety at lower costs compared to more advanced models. However, they are heavier than Types III and IV, which may restrict their use in applications where weight is critical.

In the transportation sector, these tanks are used to power fuel cell vehicles such as buses, trucks, and utility vehicles. In industry, they play a crucial role in the production of chemicals such as NH_3 and petroleum refining. They also serve in energy storage, storing excess energy from renewable sources that can be converted back to electricity via turbines or fuel cells during periods of low demand. Furthermore, Type II tanks find their place in various industrial applications, particularly in sectors requiring mobility and energy efficiency. They are frequently employed in transportation for buses and trucks, as well as in mobile applications like portable generators. In heavy industry, they provide hydrogen for metallurgy and chemical manufacturing and are also used in R&D to test new hydrogen-related technologies. Type II tanks, combining a metal shell and composite reinforcement, offer an optimal balance between weight, cost, and performance, making them attractive for a multitude of applications requiring both robustness and lightness.

3.3.3.2.3 Hydrogen Storage Tank Type III

Type III tanks for compressed hydrogen storage represent an innovative and efficient storage solution, crucial for applications requiring both high pressure and lightness.

At the core of their design, these tanks incorporate a metallic liner, often made of aluminum or lightweight steel, which plays an essential role in ensuring the tank's sealability and initial mechanical strength. This liner is then wrapped in a robust layer of composite materials, such as carbon or glass fibers, which provide the majority of the resistance to the high internal pressure, typically ranging between 350 and 700 bars. This combination of materials allows Type III tanks to exhibit optimal strength and safety while significantly reducing their weight compared to Types I and II, which are primarily metallic and considerably heavier.

The significant weight reduction of Type III tanks makes them particularly ideal for applications where weight is a critical factor influencing energy efficiency and overall performance. For example, in the field of hydrogen vehicles, component lightening is crucial to maximize the distance traveled per unit of consumed hydrogen and to improve vehicle maneuverability. Type III tanks, with their lightweight profile, directly contribute to fuel consumption reduction and increased vehicle range, vital aspects for the competitiveness of hydrogen vehicles in the market. In addition to their role in the transportation sector, Type III tanks find applications in various industrial fields where the unique properties of hydrogen as an energy carrier are exploited. In the energy sector, for example, these tanks are used to store hydrogen produced from renewable energy sources. The stored hydrogen can be used to regenerate electricity during periods of high demand or low renewable production, thus helping to stabilize the power grid. In the aerospace sector, hydrogen stored in Type III tanks could potentially power aircraft, thereby reducing carbon emissions and increasing flight energy efficiency. Finally, in maritime applications, the use of these tanks to provide a clean energy source to ships helps reduce atmospheric pollutant emissions in port areas and on the high seas, contributing to more ecological navigation. These uses demonstrate how Type III tanks, thanks to their efficiency and adaptability, play a key role in the transition to more sustainable and environmentally friendly energy systems.

3.3.3.2.4 *Hydrogen Storage Tank Type IV*

Type IV tanks for compressed hydrogen storage are at the forefront of technology, designed for maximum efficiency while minimizing weight. These tanks consist primarily of two elements: a polymer liner and an outer shell. The polymer liner, made of materials like polyethylene or polyamide, serves a critical role in maintaining the tank's seal, preventing high-pressure hydrogen gas from escaping. It offers chemical compatibility, mechanical strength, and resistance to extreme pressures and temperatures. Manufacturing techniques such as rotational molding and extrusion-blow molding ensure uniformity and reliability in production.

High-density polyethylene (HDPE) and polyamides like PA11 and PA12 are chosen for their specific properties, offering chemical resistance, flexibility, and durability. The choice between rotational molding and extrusion-blow molding depends on factors such as cost, quality control, and production volume. Overall, Type IV tanks represent a cutting-edge solution for hydrogen storage, combining advanced materials and manufacturing techniques to meet stringent performance and safety requirements.

Type IV tanks feature an advanced composite envelope crucial for internal pressure resistance and overall structural strength. Reinforced by carbon or glass fibers, this composite layer is wound around the polymer liner to maximize durability and resistance under high pressures. Typically utilizing epoxy resin or thermosetting polyester, the manufacturing process creates a solid structure that securely holds the fibers in place. Carbon fibers are preferred for their strength-to-weight ratio and rigidity, while glass fibers offer affordability and mechanical resilience. These composites offer exceptional pressure resistance, crucial for high-pressure gas storage, and their lightweight nature simplifies transportation and installation. Additionally, the durability is enhanced by the use of carbon and glass fibers, effectively resisting corrosion, fatigue, and impacts. However, challenges such as the high cost of carbon fibers and the sensitivity of some composite matrices to UV and moisture necessitate protective coatings or special formulations to prevent degradation. Despite these challenges, ongoing research and innovation in composite materials and manufacturing processes hold promise for overcoming these obstacles in the future.

Type IV tanks represent the cutting edge in hydrogen storage technology, offering significant advantages in terms of lightweight construction, efficiency, and safety. They are particularly well-suited for hydrogen-powered vehicles where weight reduction is critical, contributing to improved overall vehicle efficiency and extended range. Additionally, their design and materials ensure excellent safety, capable of handling significant pressure and temperature variations without risk of rupture. While more expensive, Type IV tanks offer superior performance in terms of weight and safety, playing a crucial role in the adoption of hydrogen as a clean energy source, especially in the mobility sector. These tanks signify a major advancement in hydrogen storage technology, essential for the future development of renewable energies.

To summarize, in the 1960s and 1970s, the use of steel and aluminum cylinders for hydrogen storage marked the earliest practical approach for medium-scale storage. Over time, fully wrapped composite tanks emerged, attracting attention due to their reduced weight and improved mobility. More recently, there has been a growing interest in hydrogen storage solutions featuring metallic and polymeric liners reinforced with fiber resin. The primary goal across all types of pressure vessels is to reduce weight by substituting metals with lighter materials while simultaneously enhancing the vessel's suitability for hydrogen storage in various ways which is why the four mentioned types of storage tanks, Types I, II, III, and IV, are now commercially available. To have a better point of view regarding the different characteristics of these types of tanks, the applied material, their specifications, and the cost of each hydrogen storage tank are shown in Table 3.6.

3.3.3.3 Comparison Between Compressed Hydrogen Storage and Other Methods

Compressed hydrogen storage stands as a prominent method in the landscape of hydrogen storage techniques, yet it necessitates a nuanced consideration alongside alternative methodologies such as liquid storage or employment of metal hydrides. Each approach exhibits unique attributes, prominently impacting factors such as energy density, economic feasibility, and infrastructural demands. Compressed storage, characterized by its simplicity and compatibility with existing infrastructure, remains a stalwart choice within various applications. However, its energy efficiency

TABLE 3.6
Specification, Material, and Cost of Various Hydrogen Cylinder Types

Type	Schematic	Material/Feature	Specification	Cost[a]
Type I	 	Steels or aluminum; heavy and high potential for internal corrosion	1460×230 mm 65 kg 175 bar 7.21 m^3 Gravimetric density 1.7%	77.5 €/kg H_2
Type II	 	Aluminum hoop-wrap, Aluminum alloy liner and glass-, aramid-, or carbon-fiber composites/heavy and high internal corrosion	394×100 mm 65 kg 175 bar 7.21 m^3 Gravimetric density 2.1%	80 €/kg H_2
Type III	 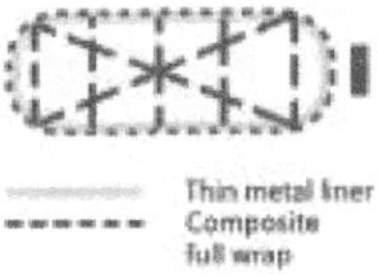	Aluminum alloy liner and glass-, aramid-, or carbon-fiber composites/relatively lightweight, resistant to corrosion	700 barL:1161 D:332 W: 61 kg H_2: 2.15 kg Gravimetric density 4.2%	530 €/kg 248 €/kg H_2
Type IV	 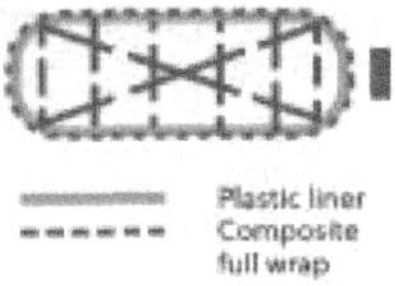	Polyethylene- or polyamide-based polymer liner and glass-, aramid-, or carbon-fiber composites/ eliminates the need for a separate metallic liner, lightest hydrogen storage systems available, equipped with safety features such as pressure relief devices and burst discs	L:1270 mm W:520 mm H: 300 mm 24.8 kg hydrogen Gravimetric density 5.7%	633 €/kg H_2

Note: (Re-used from ref. [1].)

[a] The cost reflects the price of the tanks capable of storing 1 kg of hydrogen, including both the tank and the hydrogen content itself.

pales in comparison to liquefaction, where the process of converting hydrogen gas into liquid form enables a higher energy density per unit volume. This elevated energy density renders liquefied hydrogen storage particularly advantageous in scenarios where space constraints or long-term storage efficiency are paramount concerns.

Conversely, metal hydrides represent another intriguing avenue for hydrogen storage, leveraging the chemical affinity between certain metals and hydrogen to form stable compounds. This approach offers potential advantages in terms of safety and low-pressure hydrogen release, which could be pivotal in applications requiring controlled hydrogen delivery or sensitive environments. However, metal hydrides are hindered by higher material costs and slower kinetics during the charging and discharging processes, factors that necessitate careful consideration in evaluating their feasibility within specific contexts. Despite these limitations, ongoing research into novel materials and optimized hydride systems hold promise for mitigating these challenges and unlocking the full potential of metal hydride storage solutions.

In navigating the diverse landscape of hydrogen storage technologies, it becomes imperative to conduct a comprehensive techno-economic analysis to discern the most suitable approach for a given application. Factors such as initial capital investment, operational costs, and system scalability must be weighed alongside performance metrics like energy density and system efficiency. Furthermore, considerations regarding safety, environmental impact, and regulatory compliance add layers of complexity to the decision-making process. As such, a multidisciplinary approach, encompassing materials science, chemical engineering, and economics, is essential in elucidating the trade-offs inherent in different storage methodologies and guiding informed decision-making toward achieving the overarching goal of a sustainable hydrogen economy.

Ultimately, the choice of hydrogen storage method hinges on a delicate balance of technical feasibility, economic viability, and environmental sustainability. While compressed storage offers simplicity and infrastructure compatibility, liquid storage boasts superior energy density, and metal hydrides hold promise for enhanced safety and controlled hydrogen release. By leveraging synergies across these diverse approaches and fostering continued innovation, the hydrogen storage landscape can evolve to meet the evolving needs of various industries and contribute to the realization of a cleaner, more sustainable energy future.

3.3.3.4 Risks, Performance, and Challenges of Compressed Hydrogen Storage

Compressed hydrogen storage necessitates meticulous risk management due to its high flammability and the high pressures involved. International standards, such as ISO 19880-1:2019 and NFPA 2, play a crucial role in ensuring safety through detailed guidelines covering tank design, installation, and operation. Despite these precautions, challenges such as hydrogen embrittlement and explosion risks remain significant issues, directly impacting the performance and energy efficiency of storage systems. Technical challenges also include energy consumption during compression and the associated operating and maintenance costs.

The scientific intricacies of compressed hydrogen storage extend beyond safety considerations to encompass material science, thermodynamics, and mechanical engineering. Hydrogen embrittlement, a phenomenon where hydrogen atoms diffuse into metal structures, weakening them over time, poses a formidable challenge in tank design and durability. Mitigating this risk requires a deep understanding of material behavior at the molecular level and the development of advanced alloys resistant to hydrogen-induced degradation. Moreover, the compression process itself involves thermodynamic considerations, as energy must be expended to overcome the repulsive forces between hydrogen molecules, leading to heat generation and efficiency losses. Optimizing compression techniques and exploring alternative storage materials are avenues of research aimed at enhancing the efficiency and viability of compressed hydrogen storage systems.

In addition to the technical challenges, the economic feasibility of compressed hydrogen storage is a critical aspect that demands attention. The upfront capital investment required for infrastructure development, including high-pressure tanks and compression facilities, often presents a barrier to widespread adoption. Furthermore, the ongoing operating and maintenance costs, coupled with the energy consumption inherent in compression, impact the overall cost-effectiveness of the storage solution. Cost–benefit analyses, informed by rigorous scientific and engineering assessments, are essential in evaluating the economic viability of compressed hydrogen storage systems and informing strategic decision-making for future hydrogen infrastructure development.

Despite the formidable challenges associated with compressed hydrogen storage, ongoing research and innovation hold promise for overcoming technical, safety, and economic hurdles. Advances in material science, thermodynamics, and compression technologies offer avenues for enhancing the efficiency, safety, and cost-effectiveness of storage systems. Moreover, synergies between scientific research, regulatory frameworks, and industry collaboration are essential for driving progress toward realizing the full potential of hydrogen as a clean and sustainable energy carrier. By addressing scientific challenges and fostering interdisciplinary cooperation, compressed hydrogen storage can play a pivotal role in the transition toward a low-carbon energy future.

3.3.3.5 Research, Innovation, and Perspectives in Compressed Hydrogen Storage

Research and innovation in the field of compressed hydrogen storage are pivotal for overcoming existing challenges and enhancing energy efficiency and safety. Current efforts are focused on developing new compression technologies that minimize energy consumption and optimize safety. Future prospects include the integration of advanced materials to reduce embrittlement and the improvement of leak detection systems. With the rapid evolution of the hydrogen industry, adaptations to new standards and regulatory requirements will continue to shape the development of more sustainable and economically viable storage solutions, thereby supporting the transition toward a hydrogen-based economy.

Scientific endeavors in compressed hydrogen storage span a broad spectrum of disciplines, encompassing materials science, thermodynamics, fluid dynamics, and safety engineering. Innovations in compression techniques, such as advanced piston designs or novel compression algorithms, aim to enhance the efficiency and reliability of hydrogen storage systems while reducing their environmental footprint. Concurrently, research into material science explores the development of hydrogen-resistant alloys and composite materials capable of withstanding high pressures and minimizing the risk of hydrogen embrittlement. These advancements are essential for addressing the technical challenges associated with compressed hydrogen storage and unlocking its full potential as a clean and sustainable energy carrier.

Moreover, advancements in leak detection technologies play a crucial role in ensuring the safety and integrity of compressed hydrogen storage systems. Novel sensor technologies, such as optical fiber-based sensors or hydrogen-specific gas detectors, offer improved sensitivity and reliability in detecting and localizing potential leaks. Coupled with advancements in data analytics and predictive maintenance, these technologies enable proactive risk management and early intervention, mitigating the potential consequences of hydrogen release incidents. As the hydrogen industry continues to expand, investments in R&D will be essential for driving innovation and realizing the promise of compressed hydrogen storage as a key enabler of the transition toward a low-carbon energy future.

REFERENCES

1. Mehr, A.S., et al., *Recent challenges and development of technical and technoeconomic aspects for hydrogen storage, insights at different scales; A state of art review.* International Journal of Hydrogen Energy, 2024. **70**: p. 786–815.
2. Osman, A.I., et al., *Advances in hydrogen storage materials: harnessing innovative technology, from machine learning to computational chemistry, for energy storage solutions.* International Journal of Hydrogen Energy, 2024.
3. Kim, D.W., et al., *Fine-tuned MOF-74 type variants with open metal sites for high volumetric hydrogen storage at near-ambient temperature.* Chemical Engineering Journal, 2024. **489**: p. 151500.

4 Hydrogen Application

4.1 GENERAL CONTEXT: APPLICATION OF HYDROGEN IN VARIOUS INDUSTRIES

Hydrogen, with its significant energy potential and minimal environmental impact when produced sustainably, has seen its use evolve substantially over time in various industrial sectors. It serves as an energy carrier for storing and transporting energy and is used in fuel cells to produce electricity, a key technology for hydrogen electric vehicles and stationary energy storage applications. Initially, hydrogen was adopted in the chemical industry in the early 20th century, primarily for ammonia synthesis. The Haber–Bosch process, developed in 1913, alone consumes about 1% of the world's energy and produces over 150 million tons of ammonia annually, using hydrogen as the main reactant. In the refining sector, since the 1950s, hydrogen has been used to improve fuel quality by removing sulfur and reducing the viscosity of petroleum products. Currently, the refining sector remains the largest consumer of hydrogen, using about 50% of the hydrogen produced globally to process approximately 700 million tons of oil per year. The fertilizer sector, primarily for ammonia production, uses about 40% of the hydrogen produced, highlighting its crucial role in modern agriculture. Additionally, the methanol industry, which consumes about 8% of the hydrogen produced, uses this hydrogen to manufacture about 75 million tons of methanol annually, serving as the base for various products like plastics and adhesives.

In the future, hydrogen usage is projected to diversify and expand significantly. For example, the H2FUTURE project in Europe aims to use green hydrogen produced by electrolysis with renewable energy to reduce CO_2 emissions from steel production. The steel industry, which accounts for about 7% of global CO_2 emissions, could potentially reduce its emissions by up to 95% by using hydrogen as a reductant instead of coal. Furthermore, fuel cell vehicles, which run on hydrogen, could increase in number from a few thousand currently to several million by 2030, promoting cleaner mobility.

As we approach 2050, renewable hydrogen will be increasingly deployed on a large scale, accompanying the expansion of new renewable power generation. This growth will be driven by advancements in technology and decreasing production costs. The integration of hydrogen into the global energy transition can be broken down into three main stages: production, transport and storage, and end-use, as indicated in Figure 4.1. For hydrogen to be viable, these stages must operate seamlessly within a well-structured roadmap. The presented schematic shows that hydrogen can serve various end-users, including thermal, electrical, and chemical sectors. Therefore, it is essential to ensure a stable hydrogen supply to meet demand across these sectors. This underscores the necessity for robust hydrogen storage plans that complement continuous hydrogen production, guaranteeing a reliable supply for all end-users.

DOI: 10.1201/9781032718453-4

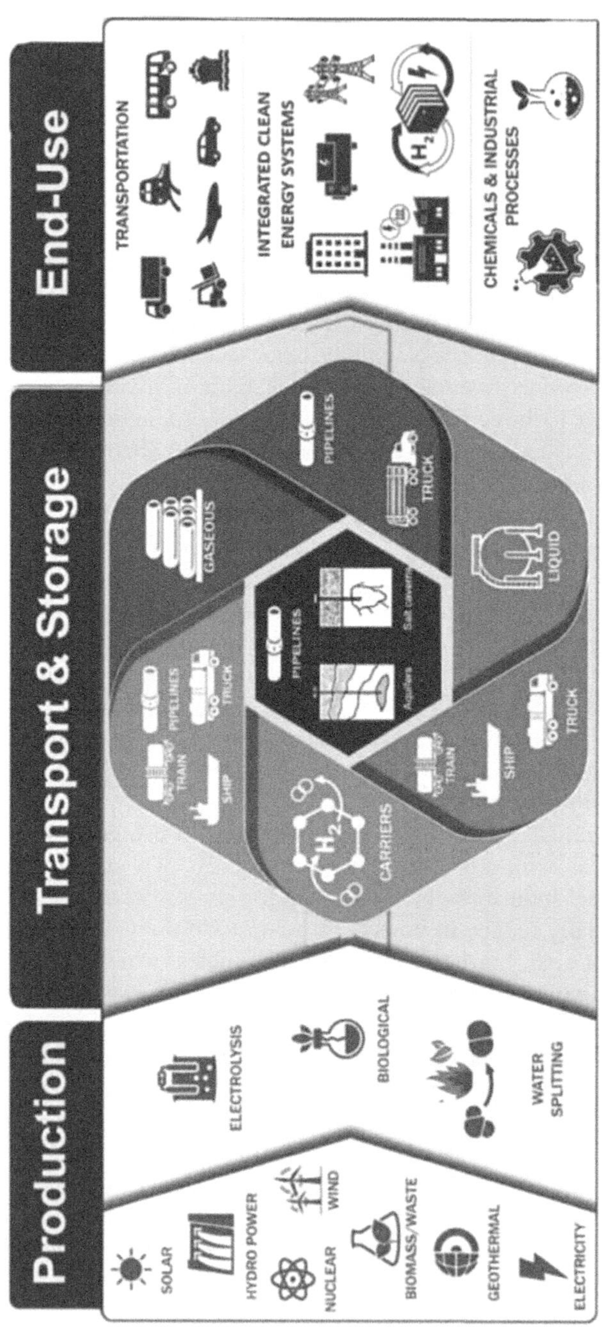

FIGURE 4.1　Basic stages for hydrogen roadmap. (Adapted from [1].)

To understand the use of hydrogen as an energy carrier in vehicles, it is essential to examine two key components: compressed hydrogen tanks and fuel cells. Hydrogen is stored in compressed form in high-pressure tanks, typically between 350 and 700 bars, in hydrogen vehicles. These tanks are made from advanced composites such as carbon fiber, which provide the necessary strength to handle the high pressure and flammability of hydrogen while remaining lightweight. Fuel cells, on the other hand, convert the chemical energy of hydrogen into electricity through a process called reverse electrolysis. Hydrogen is introduced at the anode, where it splits into hydrogen ions and electrons. The electrons produce an electric current that powers the vehicle's motor, while the ions pass through a membrane to the cathode, where combined with oxygen from the air, they form water. This process makes fuel cell vehicles an environmentally friendly solution, with their only by-product being water vapor. The advantages of hydrogen electric vehicles include zero emissions, a driving range comparable to gasoline vehicles, and quick refueling. However, this technology still faces several challenges before it can become widespread. These challenges include the still high cost of the technology, insufficient refueling infrastructure, and hydrogen production that largely relies on fossil fuels, although more eco-friendly methods like electrolysis do exist.

These figures and examples show how hydrogen, from its humble beginnings as a chemical aid, is becoming a cornerstone for a future economy based on clean and renewable energies. In this chapter, we explore the main industrial sectors where hydrogen is employed, focusing particularly on its use in transportation. We examine in detail the Type IV compressed hydrogen tank and the fuel cell, two key innovations that facilitate the integration of hydrogen as a clean energy source in vehicles. These technologies are essential for understanding the role of hydrogen in transitioning to more sustainable transportation methods. Thus, while promising a sustainable future, hydrogen vehicle technology requires technological advancements and substantial investments to become a viable alternative to fossil fuels.

4.2 OIL REFINING

Hydrogen plays an essential role in oil refining, particularly due to its ability to improve the quality of finished fuels and meet stringent environmental standards. The hydrocracking process (Figure 4.2) is a perfect illustration of this application. In this process, hydrogen is injected into a reactor with crude oil or its heavier fractions. The mixture is then exposed to high temperatures, typically between 260°C and 425°C, and pressures of up to 200 bars, promoting the breaking of chemical bonds in the hydrocarbons. A catalyst, often based on platinum or molybdenum, accelerates the reaction, facilitating the decomposition of long hydrocarbon chains into smaller structures while integrating hydrogen atoms. Hydrocracking offers several significant advantages, including increasing the yield of light fuels such as gasoline and diesel, which are more in demand on the market. This process also improves fuel quality, producing substances with a lower boiling point and better stability, making them more efficient and less polluting. Additionally, hydrocracking can process a wide range of crude oils, including those that are heavier and more sulfurous, transforming these difficult inputs into valuable products.

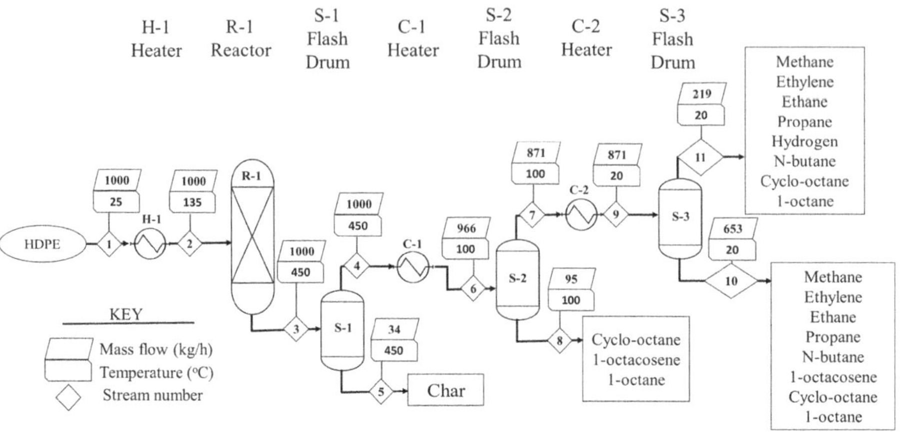

FIGURE 4.2 Hydrocracking in the oil industry.

(Adapted from [2].)

A particularly notable example of the use of hydrogen in refining is the production of ultra-low sulfur diesel. This diesel, produced by hydrocracking, not only breaks down long hydrocarbon chains but also removes sulfur from fuel components. This step is crucial for meeting strict standards that limit sulfur emissions from vehicles, significantly reducing air pollution. In summary, hydrocracking, made possible by the use of hydrogen, is indispensable to modern oil refining as it allows for the production of cleaner and more efficient fuels while optimizing the use of petroleum resources.

4.3 AMMONIA PRODUCTION

Ammonia is an essential molecule in the chemical industry, ranking as the second most produced substance by volume each year. Its applications are varied, primarily in the agricultural sector where it is used to manufacture fertilizers, but also in the energy sector where it has the potential to serve as a clean fuel. Liquid ammonia, in particular, is being considered as an alternative to fossil LPG, offering the advantage of not emitting CO_2 during combustion, making it attractive for maritime and aviation transport as a potential replacement for heavy fuels and kerosene. However, although ammonia presents these promising advantages, it is currently produced mainly from fossil resources, raising questions about its viability as a sustainable solution. These concerns include the technical feasibility of producing ammonia from renewable energy sources and its potential role as the "fuel of the future."

The industrial synthesis of ammonia is dominated by the Haber–Bosch process (as illustrated schematically in Figure 4.3), which combines hydrogen with nitrogen to produce ammonia. Hydrogen is typically produced by steam reforming of natural

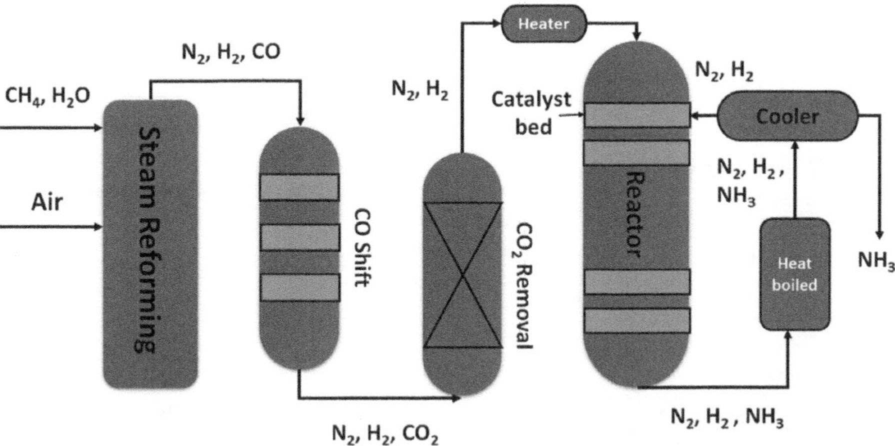

FIGURE 4.3 The diagram of the Haber–Bosch process.

(Adapted from [3].)

gas, a reaction in which methane reacts with water at high temperatures, producing CO and hydrogen [3]:

$$CH_4 + H_2O \rightarrow CO + 3H_2$$

This reaction is followed by the WGS reaction, where CO is converted to CO_2 while producing more hydrogen:

$$CO + H_2O \rightarrow CO_2 + H_2$$

This hydrogen is then used to react with nitrogen, under high pressure and temperature conditions in the presence of an iron-based catalyst to form ammonia through an exothermic and reversible reaction:

$$N_2 + 3H_2 \rightarrow 2NH_3$$

The large-scale production of ammonia has a profound impact not only on agriculture, by providing essential fertilizers such as urea and ammonium nitrate, but also on the maritime and aviation sectors as an alternative fuel. Thus, the diverse uses of ammonia, ranging from fertilizers to energy, underscore its strategic importance and potential in the transition toward more sustainable industrial practices.

4.4 METHANOL PRODUCTION

Hydrogen is a crucial element in the production of methanol, a compound widely used as a solvent, antifreeze, fuel, and as a precursor in the synthesis of many chemicals. The predominant industrial process for methanol production involves the

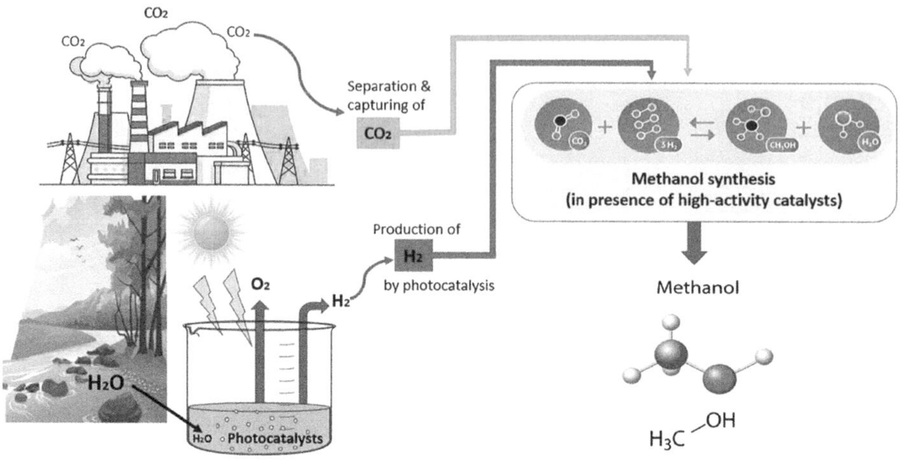

FIGURE 4.4 Methanol production via the traditional route over Cu/ZnO catalyst. **(Adapted from [4].)**

reaction of CO and hydrogen (H_2) in the presence of a copper-based catalyst, known as methanol synthesis (Figure 4.4). Methane, the main component of natural gas, is converted into hydrogen through steam reforming, a reaction where methane reacts with steam at high temperatures to form CO and hydrogen:

$$CH_4 + H_2O \rightarrow CO + 3H_2$$

This reaction is followed by the WGS reaction, where the CO produced reacts with more steam to generate more hydrogen and CO_2:

$$CO + H_2O \rightarrow CO_2 + H_2$$

Once the hydrogen is obtained, it is combined with CO in a reactor under high pressure and temperature in the presence of a copper catalyst to produce methanol:

$$CO + 2H_2 \rightarrow CH_3OH$$

CO_2 can also serve as a reactant in a reaction that also produces methanol and water:

$$CO_2 + 3H_2 \rightarrow CH_3OH + H_2O$$

Methanol is essential in several industries. It is used as an antifreeze, solvent, and fuel, and serves as an intermediate in the production of substances such as formaldehyde and acetic acid. A particularly innovative application of methanol is in direct methanol fuel cells (DMFC), which use liquid methanol to generate electricity. This technology offers advantages for energy storage and distribution, crucial for portable applications and electric vehicles where compact energy storage is needed. Accordingly, hydrogen used in methanol synthesis not only demonstrates the versatility of this chemical element but also its strategic role in providing sustainable energy solutions and reducing dependence on fossil fuels.

4.5 METALLURGICAL INDUSTRY

In the metallurgical industry, hydrogen is primarily used as a reducing agent to extract pure metals from their ores, notably in the production of tungsten and molybdenum. This role of hydrogen is essential for obtaining high-purity metals crucial for various industrial and technological applications. Hydrogen reduces metal oxides to pure metal through a high-temperature process. For instance, in tungsten production, tungsten oxide (WO_3) is transformed into metallic tungsten using hydrogen in the following reaction:

$$WO_3 + 3H_2 \rightarrow W + 3H_2O$$

This reaction occurs at temperatures between 700 and 1000°C, where hydrogen efficiently reduces WO_3, releasing water as a by-product.

Similarly, molybdenum is produced by reducing molybdenum oxide (MoO_3) with hydrogen, as described by the chemical reaction:

$$MoO_3 + 3H_2 \rightarrow Mo + 3H_2O$$

Like tungsten, this reaction requires high temperatures to ensure effective reduction and produce pure metallic molybdenum, with water as the only residue. The use of hydrogen to reduce metal oxides to pure metals is particularly advantageous in sectors where metal purity is crucial for their properties and performance. Tungsten, for example, is used in filament production for bulbs and components in the electrical industry, while molybdenum finds applications in manufacturing corrosion-resistant alloys and as a catalyst in the chemical industry. In conclusion, hydrogen, as a reducing agent in metallurgy, not only enables the production of high-purity metals but also contributes to cleaner and more sustainable manufacturing processes. Hydrogen reduction of metal oxides is an effective method to minimize impurities and reduce greenhouse gas emissions, aligning with the industry's sustainable development goals.

4.6 ELECTRONICS INDUSTRY

Hydrogen plays a very important role in the electronics industry, particularly in the manufacturing of semiconductors and other essential electronic components. This gas is primarily used for its reducing properties, which are indispensable in several key stages of integrated circuit production. One of the primary uses of hydrogen in this field is creating a reducing atmosphere during thin film deposition and etching processes. Reducing atmospheres are necessary to prevent oxidation of metal surfaces during these processes, which is crucial for maintaining the purity and functionality of electronic components. For example, in the production of silicon wafers, hydrogen is used to reduce metal oxides and purify silicon by removing impurities such as phosphorus and boron.

A significant example of hydrogen use is in the chemical vapor deposition (CVD) process for manufacturing silicon films. During CVD, silicon-containing gas precursors are introduced into a reaction chamber where they are decomposed at high temperatures in the presence of hydrogen. Hydrogen acts here as a reducing agent, preventing the formation of undesirable oxides on the silicon substrate. The reaction can be represented as follows:

$$SiH_4(silane) + 2H_2 \rightarrow Si + 4H_2$$

In this reaction, silane (SiH_4) decomposes under heat, and the additional hydrogen helps stabilize the environment by consuming any residual oxygen, forming water vapor that is then removed from the system. This process ensures the growth of a pure and uniform silicon film on the substrate, which is essential for the performance of final electronic devices. Presumably, hydrogen is indispensable in the electronics industry for its reducing properties, facilitating the manufacture of high-purity and high-performance semiconductors that are essential in modern technology. Its ability to create controlled atmospheres conducive to chemical reduction of oxides and metal purification is a major asset in the field of electronic device manufacturing.

4.7 FOOD INDUSTRY

In the food industry, hydrogen finds a specific yet important application: the hydrogenation of vegetable oils. This process involves converting liquid oils into solid fats, which is essential for the production of many food products such as margarine, pastries, and certain types of snacks. Hydrogenation of vegetable oils not only modifies the texture and stability of food products but also improves their shelf life. Hydrogenation is achieved by adding hydrogen atoms to the double bonds of unsaturated fatty acids present in vegetable oils. This process is typically catalyzed by a metal, usually nickel, which facilitates the reaction of oil molecules with hydrogen. During this reaction, vegetable oils, primarily triglycerides (composed of glycerol and fatty acids), are exposed to hydrogen in the presence of the catalyst at high temperatures and pressure. The hydrogenation reaction can be represented by the following chemical equation for a generic unsaturated fatty acid:

$$R - CH = CH - R' + H_2 \rightarrow R - CH_2 - CH_2 - R'$$

In this equation, R and R' represent the carbon chains of the fatty acid, and H_2 represents hydrogen. The double bond in the fatty acid (=) is converted into a single bond (single bonds between carbon atoms), while hydrogen atoms are added to each carbon of the former double bond. This process can be partial, resulting in partially hydrogenated oil that contains both saturated fatty acids (without double bonds) and unsaturated fatty acids (with one or more double bonds), or complete, yielding fully saturated fat. Partially hydrogenated fats may contain trans fatty acids, which have been associated with various health issues, leading to stricter regulations and reformulations in the food industry. Despite these concerns, hydrogenation remains a crucial technique for producing certain types of edible fats, playing a key role in the texture, flavor, and stability of finished products. It also allows manufacturers to manipulate the physical properties of fats to meet specific requirements for different food products.

4.8 GLASS INDUSTRY

Hydrogen plays a crucial role in the glass industry, primarily due to its reducing properties. These properties are particularly valuable during the production of high-quality glass, where preventing the oxidation of metals and other components is crucial. Undesirable oxidation during glass melting can lead to defects in the final product, such as inclusions, streaks, or color changes, compromising its clarity and strength. In the glass manufacturing process, raw materials such as silica sand (SiO_2), soda ash (Na_2O), and lime (CaO) are melted together at extremely high temperatures to form glass. During this process, the presence of metals like iron, which may be present as impurities in silica sand, needs careful management. Iron, in particular, can cause undesirable coloration of the glass, typically tinting it green or brown depending on its oxidation state. To control the oxidation state of iron and other metals, hydrogen is introduced into the melting furnace. Hydrogen acts as a reducing agent, converting iron oxides from their Fe_2O_3 form (more oxidized) to FeO (less oxidized), which has less impact on glass color. The reaction can be represented by the following equation:

$$Fe_2O_3 + H_2 \rightarrow 2FeO + H_2O$$

This reaction demonstrates how hydrogen reduces Fe_2O_3 to FeO while producing water as a by-product, which evaporates at high temperatures. This reduction process helps maintain the transparency and purity of glass by limiting iron-induced discoloration. Thus, the use of hydrogen in glass production exemplifies its traditional application in an industrial context, contributing to the production of high-quality materials while offering the potential for more sustainable and less polluting practices. By reducing the oxidation of glass components, hydrogen not only enhances product quality but also minimizes defects, ensuring increased efficiency in glass production and reduced material wastage.

4.9 TRANSPORTATION INDUSTRY

The use of hydrogen in the transportation industry is booming, driven by the need to reduce greenhouse gas emissions and find sustainable alternatives to fossil fuels. Among the most promising applications are FCEVs (Figure 4.5). These vehicles use a fuel cell to convert stored hydrogen into electricity, which then powers an electric motor. Unlike battery electric vehicles (BEVs), FCEVs can be refueled quickly and offer a range comparable to gasoline vehicles, while only emitting water vapor, making them extremely clean. Hydrogen is also finding applications in heavy transport, including trucks, buses, and delivery fleets. It is particularly suited for long distances

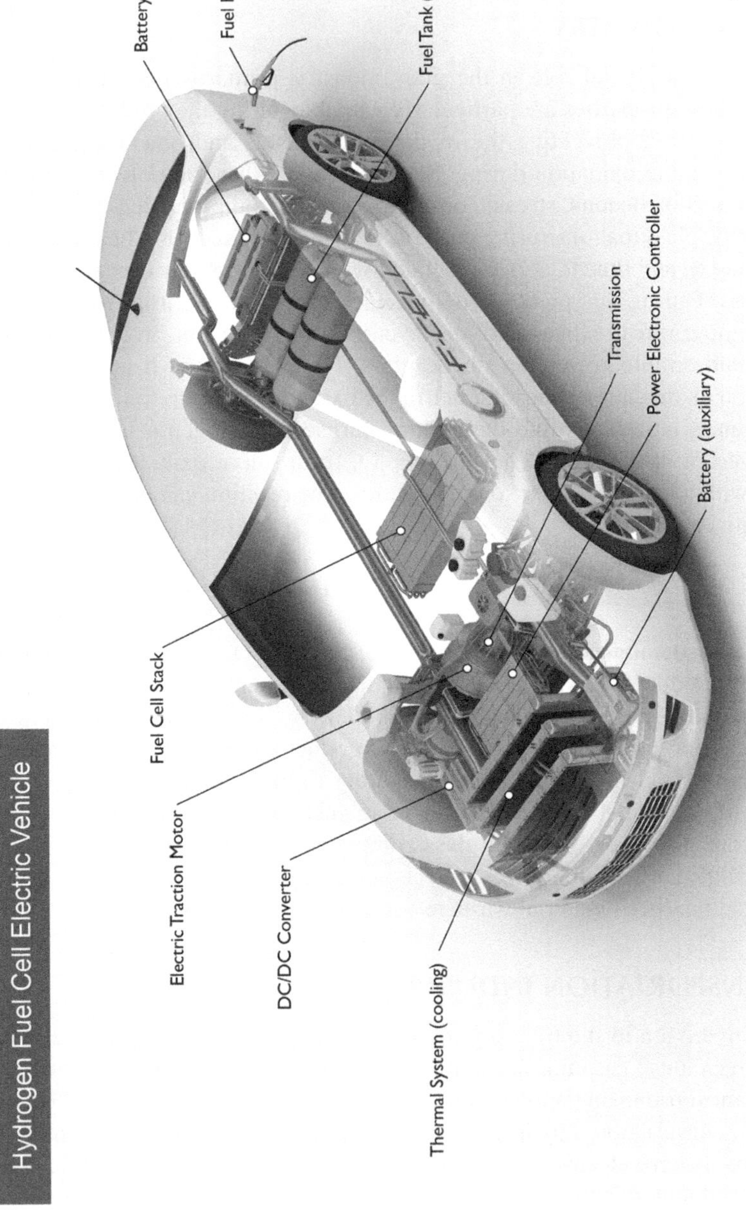

FIGURE 4.5 Different elements of a hydrogen fuel cell vehicle. (Adapted from [5].)

and heavy loads due to its high energy density. In aviation and the maritime sector, hydrogen is considered a potential fuel to reduce emissions of pollutants and GHGs, although these applications are still at an experimental stage.

The development of appropriate refueling infrastructure is essential to support the adoption of hydrogen in transport. This includes the construction of hydrogen refueling stations capable of providing fuel quickly and safely. However, the adoption of hydrogen as a solution for sustainable mobility faces several challenges, including the high cost of technologies, the predominantly fossil-based production of hydrogen, and the logistical challenges associated with its transport and large-scale storage. The long-term viability of hydrogen in transport will depend on resolving these issues and establishing a robust infrastructure.

4.9.1 Hydrogen Mobility

4.9.1.1 Definition and Functioning

Hydrogen mobility refers to the use of hydrogen as a fuel source for vehicles. This concept includes a wide range of transportation modes, such as cars, buses, trucks, and trains, which utilize hydrogen fuel cells or hydrogen combustion engines. Fuel cells work by converting hydrogen and oxygen from the air into electricity, producing only water as a by-product. This electricity then powers an electric motor that propels the vehicle. Unlike fossil fuel-powered vehicles, hydrogen vehicles emit no GHGs during operation, making them environmentally beneficial.

The functioning of hydrogen mobility involves several key steps. First, hydrogen must be produced, typically through water electrolysis or natural gas reforming. Once produced, hydrogen is stored in high-pressure tanks onboard vehicles. When a vehicle is in operation, the hydrogen is directed to the fuel cell, where it reacts with oxygen to produce electricity, which then powers the vehicle's electric motor. To support this technology, a network of hydrogen refueling stations is necessary, enabling vehicles to quickly and efficiently refill their tanks. This infrastructure is crucial for facilitating the widespread adoption of hydrogen mobility and promoting a transition to cleaner and more sustainable transportation.

4.9.1.2 Types of Vehicles Using Hydrogen as Fuel

Vehicles that use hydrogen as a fuel primarily fall into two categories: FCEVs and hydrogen combustion vehicles. FCEVs, such as the Toyota Mirai and Hyundai Nexo, are the most common and operate by converting hydrogen into electricity via a fuel cell to power an electric motor. These vehicles are known for their high range and fast refueling times, similar to gasoline vehicles. Besides passenger cars, FCEVs also include urban buses, delivery trucks, and heavy-duty vehicles developed by companies like Nikola and Hyundai for long-distance freight transport.

Additionally, trains and boats are beginning to adopt hydrogen as a fuel source. For example, Alstom's Coradia iLint hydrogen train, used in Germany, offers a clean alternative to diesel trains on nonelectrified lines. Similarly, hydrogen-powered boats and ferries are emerging, particularly in Norway, where the maritime sector seeks to reduce emissions. Although still rare, hydrogen combustion vehicles represent another development avenue. These vehicles function similarly to traditional internal combustion engines but burn hydrogen instead of fossil fuels, producing mainly water vapor as emissions. These various types of vehicles demonstrate hydrogen's versatile potential to revolutionize the transportation sector across multiple applications.

4.9.1.3 Environmental Benefits

Hydrogen mobility offers several significant environmental benefits. First, hydrogen vehicles, particularly those equipped with fuel cells, emit no pollutants at the tailpipe. They only produce water vapor as a by-product, thereby eliminating emissions of CO_2, fine particulates, CO, and nitrogen oxides (NO_x), which are commonly associated with internal combustion engines. This feature is crucial for improving air quality in urban areas, where vehicle-related air pollution is a major public health concern.

Moreover, when hydrogen is produced from renewable sources, such as water electrolysis powered by solar or wind energy, the environmental impact of hydrogen mobility is further reduced. This production method, known as green hydrogen, allows for a complete decoupling of hydrogen production from fossil fuels, thereby eliminating CO_2 emissions associated with energy production. Integrating green hydrogen into the transportation sector can create a fully clean energy value chain, significantly contributing to global goals of reducing greenhouse gas emissions and combating climate change.

4.9.1.4 Infrastructure (Refueling Stations, Production, Distribution)

Supporting hydrogen mobility requires a complex and well-coordinated infrastructure encompassing production, distribution, and refueling stations. Hydrogen production can be achieved through various methods, including natural gas reforming and water electrolysis. Production facilities need to be strategically located to minimize transport costs and maximize distribution efficiency. Once produced, hydrogen must be compressed or liquefied for storage and transport. Adequate storage facilities are essential to ensure a sufficient reserve of hydrogen to meet fluctuating demand.

The distribution of hydrogen requires a network of specialized pipelines or the use of tanker trucks capable of transporting compressed or liquid hydrogen. Establishing this network is crucial to connect production centers with refueling stations. These refueling stations, similar to traditional gas stations, must be equipped with high-pressure pumps to fill hydrogen vehicle tanks quickly and safely. The density of this refueling network must be sufficient to ensure convenience for hydrogen vehicle users, thereby encouraging their adoption. Significant investments and collaboration between the public and private sectors are necessary to develop this infrastructure effectively and sustainably.

4.9.1.5 Key Industrial Players

The key industrial players involved in developing hydrogen mobility include automotive manufacturers, energy production companies, and fuel cell technology providers. Among automotive manufacturers, leaders like Toyota, Hyundai, and Honda have pioneered the commercialization of hydrogen vehicles, with iconic models such as the Toyota Mirai and Hyundai Nexo. These companies invest heavily in R&D to improve fuel cell technology's efficiency and reduce costs. Other major manufacturers, such as BMW and Daimler, are also engaged in developing hydrogen vehicles, particularly exploring applications for trucks and buses.

In the energy production sector, companies like Air Liquide, Linde, and Shell play crucial roles in hydrogen production, storage, and distribution. These companies are developing essential infrastructure, such as hydrogen production plants via electrolysis and networks of hydrogen refueling stations. Additionally, specialized fuel cell technology companies, such as Ballard Power Systems and Plug Power, provide the necessary components to equip vehicles and other mobility applications. These players often collaborate with governments and research institutions to accelerate innovation and overcome the technical and economic challenges associated with hydrogen mobility.

4.9.1.6 Comparison with Other Clean Transport Technologies (Electric and Biofuels)

Hydrogen mobility offers several distinct advantages over other clean transport technologies, such as BEVs and biofuel-powered vehicles. One of the main advantages of hydrogen vehicles is their fast refueling time, similar to gasoline vehicles, unlike electric vehicles, which generally require longer charging times. Moreover, hydrogen vehicles offer a comparable range to traditional vehicles, making them particularly attractive for long trips and commercial applications like trucks and buses. Additionally, producing hydrogen from renewable sources can potentially reduce greenhouse gas emissions over the fuel's entire lifecycle.

In contrast, BEVs benefit from an increasingly developed charging infrastructure and superior energy efficiency, as they do not need to convert another type of fuel into electricity to operate. BEVs are also generally simpler in terms of technology and maintenance, potentially resulting in lower long-term operating costs. Biofuels, meanwhile, represent an interesting intermediate solution, allowing the use of existing fuel distribution infrastructure while reducing CO_2 emissions compared to traditional fossil fuels. However, their production can be limited by the availability of raw materials and the environmental impacts associated with cultivating these materials. Each technology presents specific advantages and unique challenges, and their complementary use could offer a more effective global solution for decarbonizing the transport sector.

4.9.1.7 Current Examples of Cities or Countries Using Hydrogen Mobility on a Large Scale

Notable examples of cities and countries using hydrogen mobility on a large scale include Japan, Germany, and South Korea. Japan, in particular, is a global leader in hydrogen. The Japanese government has invested heavily in hydrogen infrastructure, with ambitious projects like the creation of a "Hydrogen Society" aiming to integrate hydrogen into various sectors, including transportation. Tokyo boasts several hydrogen refueling stations and uses hydrogen buses for public transport, particularly visible during the 2021 Olympics. Additionally, Japanese companies like Toyota play a crucial role in promoting hydrogen vehicles, with models like the Mirai becoming emblematic.

Germany is another pioneer in adopting hydrogen mobility. The country has established a network of hydrogen refueling stations through the H_2 Mobility program, facilitating the use of hydrogen vehicles. Germany has also launched initiatives to integrate hydrogen into rail transport, with the deployment of hydrogen trains like Alstom's Coradia iLint operating in Lower Saxony. In South Korea, the government has announced ambitious plans to become a "hydrogen society" with the goal of circulating 200,000 hydrogen vehicles and building 450 refueling stations by 2025. Hyundai, a key player, has developed the hydrogen SUV Nexo and is also investing in developing hydrogen trucks for freight transport. These initiatives show how different countries and cities are integrating hydrogen into their transport systems to create more sustainable mobility solutions.

4.9.1.8   Global Deployment Scenarios for Hydrogen Mobility by 2030 and 2050

By 2030, global deployment scenarios for hydrogen mobility envisage increased adoption of hydrogen vehicles, supported by favorable government policies and infrastructure investments. Many countries plan to build thousands of hydrogen refueling stations to enhance the accessibility and convenience of these vehicles. For example, the European Union, through its "European Green Deal" initiative, aims to have a comprehensive network of hydrogen refueling stations to support hydrogen mobility. Additionally, technological advancements should significantly reduce the costs of producing green hydrogen, making this option more competitive with fossil fuels and other clean transport solutions. China, Japan, and South Korea, with their ambitious decarbonization plans, are also expected to be at the forefront of this movement, integrating fleets of commercial and public transport hydrogen vehicles.

By 2050, hydrogen mobility could play a central role in a globally decarbonized transport system. Long-term scenarios envisage a transition to hydrogen production almost entirely based on renewable sources, with sophisticated and widely available infrastructure for hydrogen storage and distribution. Hydrogen vehicles could become common not only in public and commercial transport sectors but also among private individuals, thanks to reduced costs and ubiquitous refueling infrastructure. Additionally, hard-to-electrify sectors such as aviation and maritime transport could adopt hydrogen as a primary fuel, contributing to a drastic reduction in CO_2 emissions. This vision relies on robust international collaborations, coherent policies, and continuous technological innovations to overcome the current and future challenges of hydrogen mobility.

4.9.2   Fuel Cell

Hydrogen fuel cells represent a promising technology in the field of renewable energy, offering an efficient solution to convert compressed hydrogen into electricity. These devices operate through an electrochemical reaction that combines hydrogen with oxygen from the air to produce water, heat, and electricity, without combustion. This makes the process not only clean but also silent, a major advantage for applications in residential and urban areas. Moreover, unlike traditional combustion engines, fuel cells can convert energy with higher efficiency and without direct CO_2 emissions.

However, several challenges need to be addressed to ensure the widespread deployment of this technology. One of the main obstacles is hydrogen production, which currently heavily relies on fossil sources. Establishing a viable infrastructure for green hydrogen production through water electrolysis, powered by renewable sources, is crucial for the future of this technology. Concurrently, developing suitable refueling infrastructure remains a major challenge to facilitate public access and adoption of this technology. The costs associated with fuel cell technologies are also a significant consideration. Currently, the use of expensive catalysts such as platinum makes initial costs quite high. Continuous research to find less-expensive alternatives is therefore crucial to make fuel cells economically competitive. The sustainability and reliability of these systems are also central concerns, requiring ongoing improvements to meet the requirements of commercial and industrial applications.

In terms of safety and regulation, rigorous measures must be implemented to manage risks associated with hydrogen storage and transport. This includes solutions for safely storing hydrogen at high pressure or low temperature, ensuring adequate autonomy for users. The potential applications of hydrogen fuel cells are vast, ranging from electric vehicles offering superior range compared to battery-powered vehicles to backup power systems for critical infrastructures such as hospitals and data centers. Additionally, these systems can play a key role in integrating renewable energies, serving as storage means to balance intermittent solar and wind energy production.

To realize the full potential of hydrogen fuel cells, coordination between scientific advancements, policy support, and investments in research and infrastructure will be necessary. Only through a holistic approach can fuel cells significantly contribute to a sustainable energy future.

4.9.2.1 Different Fuel Cell Types

Fuel cells are versatile energy conversion devices categorized by their electrolyte type and operational temperature. Each type serves distinct purposes suited to specific applications. Proton Exchange Membrane Fuel Cells (PEMFC) utilize a polymer membrane electrolyte and function at lower temperatures (around 80°C). This makes them ideal for mobile applications, particularly in electric vehicles, due to their rapid start-up capability and ability to adjust power output swiftly. SOFCs employ a solid ceramic electrolyte and operate at very high temperatures (800°C to 1000°C). While highly efficient and compatible with various fuels, SOFCs are predominantly used in stationary applications owing to their elevated operational temperatures, which pose challenges for vehicle integration.

Alkaline Fuel Cells (AFCs) utilize an aqueous potassium hydroxide solution as the electrolyte and operate at moderate temperatures. Once prevalent in space missions, AFCs have become less common today due to the electrolyte's vulnerability to CO_2, limiting their widespread adoption. Phosphoric Acid Fuel Cells (PAFCs) use phosphoric acid as the electrolyte, operating at temperatures slightly higher than PEMFC (around 200°C). PAFCs are typically employed in stationary settings such as backup power systems and cogeneration units, where their reliability and efficiency are advantageous. Molten Carbonate Fuel Cells (MCFC) employ carbonate salts as the electrolyte and operate at approximately 650°C. Like SOFCs,

TABLE 4.1
Different Types of Fuel Cells

Fuel Cell Type	Operating Temperature	Electrolyte	Energy Efficiency
PEMFC (Proton Exchange Membrane Fuel Cell)	60–80°C	Polymer membrane	40–60%
AFC (Alkaline Fuel Cell)	60–90°C	Potassium hydroxide	60–70%
PAFC (Phosphoric Acid Fuel Cell)	160–220°C	Phosphoric acid	40–55%
MCFC (Molten Carbonate Fuel Cell)	600–650°C	Molten carbonate salts	45–60%
SOFC (Solid Oxide Fuel Cell)	800–1000°C	Solid ceramic (oxide)	50–60%

MCFCs are primarily utilized in stationary applications due to their high operating temperatures, which facilitate efficient energy conversion but complicate mobile deployment.

Table 4.1 compares different types of fuel cells, incorporating key data such as operating temperature, type of electrolyte used, and energy efficiency. These characteristics may vary slightly depending on specific design and application, but the figures provided here offer a good general approximation.

The explanation provided highlights the unique characteristics of each type of fuel cell, emphasizing their suitability for various applications. PEMFCs are particularly well-suited for mobile applications like vehicles due to their lower operating temperatures and rapid start-up capability. Conversely, AFCs, while highly efficient, are less commercially prevalent due to their sensitivity to CO_2 contamination, which can affect the electrolyte. PAFCs are noted for their robustness against CO_2 contamination and are typically used in medium-power stationary applications. MCFCs excel in industrial applications and cogeneration due to their ability to utilize CO_2 and operate efficiently at high temperatures, enabling effective heat recovery. SOFCs offer flexibility in fuel types and are efficient in heat recovery as well, but their high operating temperatures present challenges in terms of materials and durability.

These figures underscore how each fuel cell technology has distinct advantages that make them more or less suitable for specific applications. The choice of technology depends heavily on the intended use case, cost considerations, and the constraints of available infrastructure. In transportation sectors such as automobiles, buses, and other vehicles, PEMFCs are predominantly favored. Their ability to efficiently generate electricity from hydrogen, coupled with their responsive power dynamics and quick start-up, makes them ideal for applications requiring frequent load changes. PEMFCs thus represent a clean alternative to ICEs in FCEVs.

4.9.2.2 Operating Principle of Fuel Cells

Fuel cells using hydrogen are devices that convert the chemical energy of hydrogen into electricity through a clean and efficient electrochemical process. This process is distinctly different from combustion because it involves no flame or direct

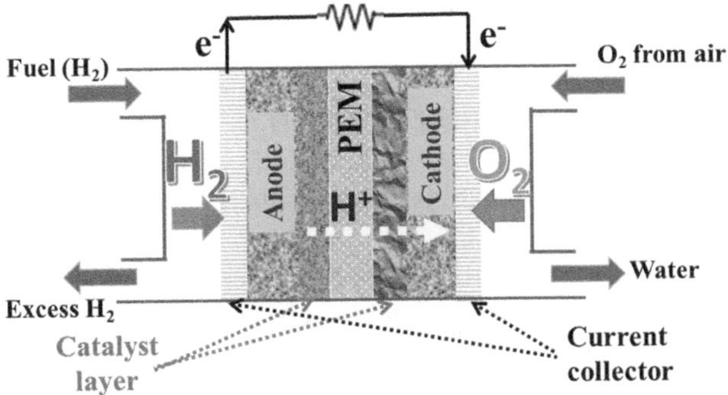

FIGURE 4.6 Operating principle of a hydrogen fuel cell.

(Adapted from [6].)

emission of atmospheric pollutants, making it particularly attractive for environmentally friendly applications. The operation of a hydrogen fuel cell begins at the anode, where hydrogen is introduced. In the fuel cell, gaseous hydrogen is dissociated into protons and electrons (Figure 4.6). This dissociation is catalyzed by a catalyst, often platinum-based, applied to the surface of the anode. The electrons released from this reaction are forced to travel through an external electrical circuit because the electrolyte in the center of the cell does not allow them to pass directly to the cathode. This flow of electrons through the external circuit generates an electrical current that can power motors, lights, or any other electrical device.

Meanwhile, the protons generated at the anode traverse the electrolyte, a membrane specially designed to allow only protons to pass while blocking electrons. Upon reaching the cathode, these protons combine with the electrons that travel through the external circuit and with oxygen from the air. This combination produces water, which is the only by-product of the reaction in a hydrogen fuel cell, along with heat. This process relies on a simple chemical reaction:

$$\text{Hydrogen} + \text{Oxygen} \rightarrow \text{Electricity} + \text{Water} + \text{Heat}$$

$$2\,H_2 + O_2 \rightarrow 2\,H_2O$$

In the fuel cell, a redox reaction occurs to create electricity and heat. At the anode, the hydrogen molecule, upon contact with a catalyst, decomposes and releases electrons that create the electrical current. This is oxidation.

$$2\,H_2 \rightarrow 4\,H^+ + 4\,e^-$$

On the other hand, at the cathode, oxygen, upon contact with the electrons released by the previous reaction, reacts. This is reduction.

$$O_2 + 4\,e^- \rightarrow 2\,O^-_2$$

Finally, the hydrogen protons, when they reach the cathode, recombine with the oxygen ions and form water.

$$4\,H^+ + 2\,O^-_2 \rightarrow 2\,H_2O$$

However, while the principle is simple, its implementation remains more complex.

This direct conversion of hydrogen into electricity without intermediate combustion makes hydrogen fuel cells particularly efficient, often more so than internal combustion engines or turbines. Moreover, since the only by-products are water and heat, fuel cells offer an extremely clean energy solution, ideal for reducing greenhouse gas emissions and combating climate change. Their ability to provide reliable and continuous energy makes them suitable for a wide range of applications, from electric vehicles to energy backup systems and distributed energy installations.

4.9.2.3 Advantages and Challenges of Fuel Cells Compared to Other Energy Sources

Fuel cells offer numerous significant advantages compared to other energy sources, particularly in terms of emissions, efficiency, and reliability. They convert the chemical energy of hydrogen into electricity through a clean and efficient electrochemical process, with water as the only by-product when operating on hydrogen. This characteristic makes them particularly attractive for contributing to greenhouse gas emission reduction and combating climate change. In terms of efficiency, fuel cells surpass traditional energy conversion methods such as internal combustion engines, which typically have energy efficiencies of only 25% to 30%. Fuel cells, on the other hand, can achieve much higher efficiencies, often exceeding 60%. Additionally, they operate quietly, making them ideal for residential applications or vehicles where noise reduction is beneficial. Versatility is another key advantage of fuel cells. They can be used in a variety of applications, from small portable devices to large energy production systems, including vehicles and backup power installations. This technology is also capable of storing excess energy produced by intermittent renewable sources like wind and solar, thereby improving the efficiency of these energy systems. The adoption of fuel cells also contributes to reducing dependence on fossil fuels, especially in transportation and energy production sectors. They offer high reliability and energy quality, crucial for critical infrastructures such as hospitals and data centers. Some fuel cells can even use various fuels, including natural gas and biogas, which can make them compatible with existing energy infrastructures.

Furthermore, the environmental impact of fuel cells is significantly reduced because they not only produce little to no direct emissions but also less thermal and noise pollution. These characteristics contribute to better air quality and a less-disrupted environment, supporting a transition to more sustainable and environmentally friendly energy systems. However, despite the many advantages, several significant challenges must be overcome to maximize their potential and facilitate widespread adoption.

One of the main challenges is the cost of fuel cells, which remains high due to the expensive materials required for their manufacture, notably platinum used as a catalyst. This constraint makes fuel cells less competitive compared to other established energy technologies and delays their integration into mass markets such as automotive and residential energy. The production and distribution of hydrogen also pose a major challenge. Currently, the majority of hydrogen is produced from fossil sources, which can offset some of the environmental benefits of fuel cells. Moreover, developing infrastructure for safe and efficient hydrogen storage and transport requires significant investments. Hydrogen refueling stations are much less common than traditional gas stations or electric charging stations, limiting the practicality of fuel cell vehicles. The durability of fuel cells is another issue. Material degradation within the cell, particularly the PEM, can reduce cell efficiency and limit its lifespan. This degradation is often accelerated by demanding operating conditions such as temperature variations and load management. Additionally, although fuel cells produce few direct emissions, the entire life cycle of hydrogen, from production to utilization, needs to be made more sustainable. Current hydrogen production methods are energy-intensive and can contribute to overall greenhouse gas emissions unless renewable energy is used for water electrolysis.

4.9.2.4 Future Prospects for Fuel Cells

The future prospects for fuel cells are highly promising, largely due to their potential to support an energy transition toward cleaner and more sustainable sources. Seen as a key technology in reducing carbon emissions, especially in the transport and industrial sectors, fuel cells could play a crucial role in the decades to come. The transportation sector, in particular, holds significant potential for fuel cell adoption, especially in light-duty vehicles, buses, and trucks. With increasing environmental regulations and pressure to reduce greenhouse gas emissions, fuel cell vehicles are viewed as a viable alternative to internal combustion engine vehicles. Several countries and regions have already begun supporting this transition through subsidies, tax incentives, and the development of infrastructure such as hydrogen refueling stations.

Beyond transportation, fuel cells find applications in residential and commercial sectors as backup power sources or cogeneration systems, which simultaneously produce electricity and heat. Their ability to provide clean and efficient energy makes them particularly suitable for eco-friendly buildings and microgrids. However, to fully realize their potential, several technological and economic challenges must be overcome. Cost reduction through technological advancements and scaling up production is crucial. Additionally, hydrogen production must become more environmentally friendly. Currently, the majority of hydrogen is produced from fossil sources, but electrolysis of water using renewable energy offers a cleaner albeit more costly alternative. The future of fuel cells will largely depend on the evolution of environmental policies, technological advancements, and market dynamics. With adequate support, these technologies can significantly contribute to a more sustainable energy future, reducing dependence on fossil fuels and decreasing global greenhouse gas emissions.

4.10 COMPRESSED HYDROGEN TANK

Hydrogen serves as an energy vector for storing and transporting energy. It is used in fuel cells to produce electricity, a key technology for hydrogen fuel cell vehicles and stationary energy storage applications. To understand the use of hydrogen as an energy vector in vehicles, it is essential to look at two key components: compressed hydrogen tanks and fuel cells. Hydrogen is stored in compressed form in high-pressure tanks, typically between 350 and 700 bars, in hydrogen vehicles. These tanks are made from advanced composites such as carbon fiber, which provides the necessary robustness to handle the high pressure and flammability of hydrogen while remaining lightweight.

Fuel cells, on the other hand, convert the chemical energy of hydrogen into electricity through a process called reverse electrolysis. Hydrogen is introduced at the anode, where it splits into hydrogen ions and electrons. The electrons produce an electric current that powers the vehicle's motor, while the ions pass through a membrane to the cathode where, with the oxygen from the air, they form water. This process makes hydrogen fuel cell vehicles an environmentally friendly solution, with water vapor being their only by-product. The advantages of hydrogen electric vehicles include zero emissions, a range comparable to gasoline vehicles, and fast refueling. However, this technology still needs to overcome several hurdles before it can be widely adopted. Among these are the high cost of the technology, insufficient refueling infrastructure, and hydrogen production that is still largely dependent on fossil fuels, although more eco-friendly methods like electrolysis are available.

Thus, while promising a sustainable future, hydrogen vehicle technology requires technological advancements and substantial investments to become a viable alternative to fossil fuels.

4.10.1 Shape and Geometry of the Tanks

In the current context of energy transition, hydrogen is emerging as a key energy vector to achieve global decarbonization goals. The efficient storage of hydrogen, particularly in compressed form, represents a major technical and scientific challenge. Among the technological solutions developed, Type IV tanks stand out for their ability to combine high performance and safety (Figure 4.7). This section is dedicated to a detailed exploration of the structure, materials, and manufacturing processes of Type IV tanks, with a particular focus on analyzing their key components: the liner, composite layer, and metal inserts. Type IV tanks represent the state-of-the-art in hydrogen storage technology, offering significant advantages in terms of lightweight construction, efficiency, and safety.

FIGURE 4.7   Hydrogen tank Type IV.

These tanks are the lightest among all types of hydrogen tanks, making them ideal for applications where weight is a major constraint, such as in hydrogen vehicles. The weight reduction contributes to improved overall efficiency of the vehicle or system where the tank is integrated. Moreover, the design and materials used in Type IV tanks ensure excellent safety, with structures capable of handling significant variations in pressure and temperature without the risk of rupture. In terms of applications, Type IV tanks are primarily used in hydrogen vehicles such as cars, buses, and trucks, where their lightweight construction is a key advantage for reducing fuel consumption and increasing range. They are also employed in stationary storage applications, where despite being more costly, their ability to minimize space and weight is crucial.

While Type IV tanks may be more expensive, they offer superior performance in terms of weight and safety, playing a pivotal role in the adoption of hydrogen as a clean energy source, especially in the mobility sectors. These tanks signify a major advancement in hydrogen storage technology, essential for the future development of renewable energies.

4.10.2 Components of Type IV Tank

A Type IV hydrogen tank is designed with essential components to ensure the safe and efficient storage of hydrogen. These tanks are primarily made from advanced composite materials, such as carbon fiber-reinforced polymers (CFRPs) or glass fiber-reinforced polymers (GFRPs), which provide mechanical strength while remaining lightweight. Inside the tank, an internal barrier is installed to prevent the diffusion of hydrogen through the tank wall, made of special polymers or multilayer coatings. The tank is also designed to securely contain hydrogen under high pressure, thanks to a robust containment system capable of withstanding high pressures without deformation or leakage. The filling and discharge of hydrogen are controlled by a dedicated valve that regulates the hydrogen flow and maintains internal pressure at a safe level. For additional safety, some tanks integrate devices like pressure relief valves or thermal fuses to prevent overpressure and overheating.

To minimize energy losses and maintain hydrogen at an appropriate temperature, thermal insulation can be applied around the tank's external surface. The attachment and support of the tank in its specific application, such as in a hydrogen vehicle or an energy storage system, are also crucial to ensure its stability and safety. These elements collectively ensure that the Type IV tank meets the strictest safety standards, capable of withstanding high pressures and varying environmental conditions while minimizing the risks of leakage or failure. In this study, we specifically focus on three key components: the liner, the composite layer, and the metal insert.

4.10.2.1 Polymer Liner

The inner layer of hydrogen tanks consists of a polymer liner, typically made from polyethylene or polyamide. This liner plays a crucial role in ensuring the tank's impermeability, preventing the high-pressure gaseous hydrogen from escaping. The primary function of this liner is to contain hydrogen under high pressures while resisting permeability. The liner performs several critical functions essential for the performance and safety of the tanks. Among these functions, ensuring impermeability is paramount as it prevents hydrogen leakage. In terms of mechanical properties, protected by a composite layer, the liner remains intact even under extremely high pressures, reaching up to 700 bars during gas loading and 1500 bars during burst tests. Despite this protection, the liner must exhibit sufficient mechanical strength to withstand specific stresses not covered by the composite layer, particularly at low temperatures, such as confined compression behavior, impact resistance, and vacuum conditions. Furthermore, the liner offers excellent chemical compatibility, acting as an effective barrier against potential chemical reactions between hydrogen and the external shell material. These characteristics, combined with the implementation processes of the polymers used, define its crucial role in optimizing the safety and efficiency of hydrogen tanks.

The polymers used for Type IV tank liners are selected based on specific properties to meet rigorous performance and safety requirements. In what follows, the most applied polymers for liner fabrication are mentioned.

4.10.2.1.1 *High-Density Polyethylene (HDPE)*

HDPE is frequently used for manufacturing liners of Type IV tanks dedicated to compressed hydrogen storage. This preference is due to HDPE's unique qualities, such as its chemical resistance, durability, and gas impermeability, which are crucial for hydrogen storage applications. Among the most advantageous properties of HDPE is its low gas permeability, essential for minimizing hydrogen leaks—a notoriously difficult gas to contain due to the small size of its molecules. In addition to its mechanical strength, which, while lower than that of some other plastics, is sufficient to withstand high internal pressures without cracking, HDPE is valued for its ductility. This ductility allows it to remain flexible even at low temperatures, reducing the risk of rupture under mechanical or thermal stress. Polyethylene is also prized for its processability, as it can be easily shaped using various techniques such as rotational molding and blow molding.

In terms of chemical resistance, HDPE withstands many chemical solvents, enabling it to contain hydrogen and other chemicals without degradation. Within a Type IV tank, the HDPE liner plays a key role in containing hydrogen and acting as a barrier against its diffusion through the tank structure. It also protects the external composite layer by preventing hydrogen from deteriorating the composite fibers and matrix. The quality of the bond between the liner and the composite layer is vital for the overall performance of the tank, as any failure at this interface can severely compromise the tank's structure. However, the use of HDPE is not without challenges. Despite its relatively low permeability, some hydrogen can still diffuse through the material, necessitating increased safety measures and monitoring to prevent the risk of gas accumulation. Additionally, the long-term behavior of HDPE under high pressure and in the presence of hydrogen is not entirely free from issues, prompting ongoing research into how these factors influence the durability and integrity of the liner. These aspects make HDPE an indispensable material in the design of Type IV tanks, contributing to an effective solution for the safe storage of high-pressure hydrogen. Table 4.2 presents some characteristics of HDPE.

TABLE 4.2

Characteristics of HDPE

Characteristic	Value
Density (g/cm^3)	0.941–0.965
Young's Modulus (MPa)	200–1600
Maximum Stress (MPa)	20–30
Maximum Strain at Ambient (%)	100–600 depending on the grade
Maximum Strain at –40°C	Flexible even at low temperatures
Permeation Coefficient to H_2 (Barrer)	0.1–1

4.10.2.1.2 Polyamide 11 (PA11)

Polyamide 11 (PA11), also known as nylon 11, is a bio-based polymer widely used in the manufacturing of liners for compressed hydrogen tanks. This material is distinguished by its unique properties, making it ideal for demanding environments such as high-pressure hydrogen storage.

PA11 offers several significant advantages for this application. First, it has excellent chemical resistance, capable of withstanding many chemicals, including hydrocarbons and solvents, which is crucial for containing gases like hydrogen. Additionally, PA11 is recognized for its effectiveness as a gas barrier, essential for minimizing hydrogen losses through diffusion across the tank liner. Its ability to provide a gas barrier is often superior to that of many other plastics, including some types of polyethylene. Furthermore, this polymer offers good impact and tensile strength, vital for tanks that must withstand high internal pressures without cracking or bursting. Unlike other polymers that can become brittle at low temperatures, PA11 maintains flexibility and ductility, reducing the risk of rupture due to temperature and pressure variations. Finally, PA11 is also valued for its good resistance to UV and abrasion, extending the lifespan of components exposed to outdoor conditions or harsh environments. In the context of compressed hydrogen tanks, PA11 is frequently chosen to manufacture the internal liner, which acts as the primary barrier to contain the gas under high pressure. These liners are typically made through processes such as extrusion or injection molding, which allow for the precise and uniform shapes needed for these applications. The excellent gas barrier capability of PA11 helps minimize hydrogen leaks, thereby enhancing the efficiency and safety of storage systems. Additionally, PA11 liners are often used in combination with carbon fiber composite wraps to create Type IV tanks, where the liner ensures impermeability and the composite provides structural strength. As a bio-based product, PA11 is also valued for its environmental attributes, offering a more sustainable alternative compared to fully petroleum-based polymers.

However, despite its many advantages, PA11 is generally more expensive than some other polymers, such as HDPE. The production of PA11 may also be less common, which could affect the availability and cost of the material for large-scale applications. PA11 is a valuable material choice for compressed hydrogen tank liners, thanks to its excellent mechanical strength, gas barrier properties, and durability,

TABLE 4.3
Characteristics of PA11

Characteristic	Value
Density (g/cm³)	1.03–1.05
Young's Modulus (MPa)	1500–1800
Maximum Stress (MPa)	45–60
Maximum Strain at Ambient (%)	50
Maximum Strain at −40°C	Performance maintained, slightly reduced
Permeation Coefficient to H_2 (Barrer)	0.1–0.3

aligned with the rigorous requirements of this application. Table 4.3 presents some characteristics of PA11.

4.10.2.1.3 Polyamide 12 (PA12)

Polyamide 12 (PA12), often referred to as Nylon 12, is a thermoplastic polymer widely used in the manufacturing of liners for compressed hydrogen tanks. While it shares some chemical and physical properties with PA11, PA12 possesses distinct characteristics that make it particularly suitable for specific applications. In terms of properties, PA12 excels in several key areas crucial for high-pressure hydrogen storage. It offers excellent chemical resistance, effectively withstanding various chemicals such as oils, greases, hydrocarbons, and solvents, which is critical for preventing chemical degradation in potentially corrosive environments. Additionally, like PA11, PA12 benefits from low gas permeability, a significant advantage for minimizing hydrogen leaks and maintaining optimal energy efficiency in tanks. Its high impact resistance and resistance to stress cracking make it the ideal material for conditions where tanks may face high mechanical stresses. It also maintains good flexibility even at low temperatures, reducing the risk of ruptures or cracks due to temperature or pressure variations, and it exhibits excellent dimensional stability, important for maintaining the structural integrity of the liner under variable conditions.

The manufacturing process of PA12 liners for hydrogen tanks typically involves advanced techniques such as extrusion or injection molding. These methods enable the production of precise and uniform liners that can then be inserted into Type IV tanks, often composed of an external shell made from composite materials. PA12 is known for its extended durability, making it a cost-effective choice for applications requiring a long lifespan. It withstands UV and other adverse environmental conditions well, making it suitable for outdoor applications. Furthermore, it bonds well with composite materials used for the outer layers of tanks, facilitating good adhesion between the liner and the composite layer.

However, similar to PA11, the cost of PA12 may be higher than that of other polymers like HDPE, and its specific processing and availability characteristics may vary depending on suppliers and product specifications. PA12 remains an excellent choice for liners in compressed hydrogen tanks, due to its combination of chemical resistance, low permeability, and durability, making it a preferred material to ensure the safety and efficiency of hydrogen storage systems. Table 4.4 presents some characteristics of PA12.

TABLE 4.4
Characteristics of PA12

Characteristic	Value
Density (g/cm³)	1.01–1.02
Young's Modulus (MPa)	1500–1800
Maximum Stress (MPa)	45–60
Maximum Strain at Ambient (%)	Environ 50
Maximum Strain at −40°C	Slightly reduced performance compared to ambient temperature
Permeation Coefficient to H_2 (Barrer)	0.05–0.1

4.10.2.1.4 Polyamide 6 (PA6)

Polyamide 6 (PA6), also known as Nylon 6, is an important member of the polyamide family and is used in a variety of industrial applications, including the manufacturing of liners for compressed hydrogen tanks. This material distinguishes itself with cost advantages and mechanical properties compared to PA11 and PA12, although it presents specific challenges, especially regarding thermal stability during molding. PA6 is valued for its high mechanical strength, rigidity, and good wear resistance, making it ideal for applications requiring high durability under mechanical stress. It is generally less expensive to produce than PA11 and PA12, making it an attractive option for large-scale applications where budget is a key consideration. However, PA6 tends to absorb more moisture than its counterparts, which can affect its mechanical and dimensional properties in humid environments. Shaping PA6 poses specific challenges due to its lower thermal stability compared to other polyamides. It requires higher processing temperatures, which can lead to material degradation if the temperature is not precisely controlled. Moreover, at high temperatures, PA6 is susceptible to hydrolytic degradation, which can compromise its mechanical properties and dimensional stability. Therefore, injection molding or extrusion techniques need to be carefully optimized to maintain polymer integrity, ensuring control over process parameters such as temperature, pressure, and cooling rate.

Despite these challenges, PA6 is used in hydrogen tank liners for its excellent mechanical properties. It is particularly suitable in situations where rigidity and strength are more critical than temperature sensitivity. To ensure liner performance and durability, stabilizing additives can be incorporated to enhance heat resistance and reduce degradation during molding and use. Liner designs can also be adapted to minimize degradation effects by reducing areas of concentrated stress and optimizing material thickness. Rigorous quality control during manufacturing is essential to ensure the liner meets the required standards for safe hydrogen storage. Although PA6 offers a cost-effective option for manufacturing hydrogen tank liners with superior mechanical properties, its sensitivity to molding conditions and thermal stability requires careful management to ensure finished products meet performance and safety requirements in compressed hydrogen storage applications. Table 4.5 presents some characteristics of PA6.

TABLE 4.5
Characteristics of PA6

Characteristic	Value
Density (g/cm^3)	1.12–1.14
Young's Modulus (MPa)	2500–3500
Maximum Stress (MPa)	75–85
Maximum Strain at Ambient (%)	60–300
Maximum Strain at −40°C	Slightly reduced performance compared to ambient temperature

4.10.2.2   Manufacturing Methods of the Liner

The manufacturing of polymer liners for Type IV tanks primarily utilizes two advanced techniques: extrusion-blow molding and rotational molding. Each method offers unique benefits and is chosen based on specific production requirements and desired properties of the final product.

4.10.2.2.1   Extrusion-Blow molding

Extrusion-blow molding is a plastic manufacturing method used to produce hollow tanks and liners, including for applications such as compressed hydrogen storage tanks. This process is particularly suitable for producing lightweight and durable parts, and it is often chosen for Type IV tank liners made of polyethylene or nylon. As illustrated in Figure 4.8, the process begins with the extrusion of a heated and melted polymer through a die to form a tubular parison, which serves as a preform resembling a soft plastic tube. The parison is then shaped within a mold designed to define the final geometry of the tank liner. Compressed air is subsequently injected into the parison, pushing it against the mold walls until it conforms precisely to the inner shape. This method ensures the liner's integrity and dimensional accuracy while optimizing material use and production efficiency.

Extrusion-blow molding offers several advantages for hydrogen tank liners. It is efficient and economical, allowing for the production of complex parts at lower costs compared to other manufacturing methods. The liners produced are lightweight yet strong, crucial for hydrogen tanks that must withstand high pressures while minimizing additional weight. This method also provides flexibility in design, accommodating complex shapes and specific contours necessary for optimal tank performance. However, challenges include material selection, critical for forming an effective barrier against hydrogen and enduring environmental and mechanical conditions within the tank. Ensuring dimensional stability during the cooling and finishing processes is also vital to maintain the liner's performance under pressure.

4.10.2.2.2   Rotational Molding

Rotational molding is a polymer processing method used to produce hollow particles ranging in capacity from 0.2 to 50,000 liters. This technique accommodates both small and large production runs, with the primary advantage being significantly

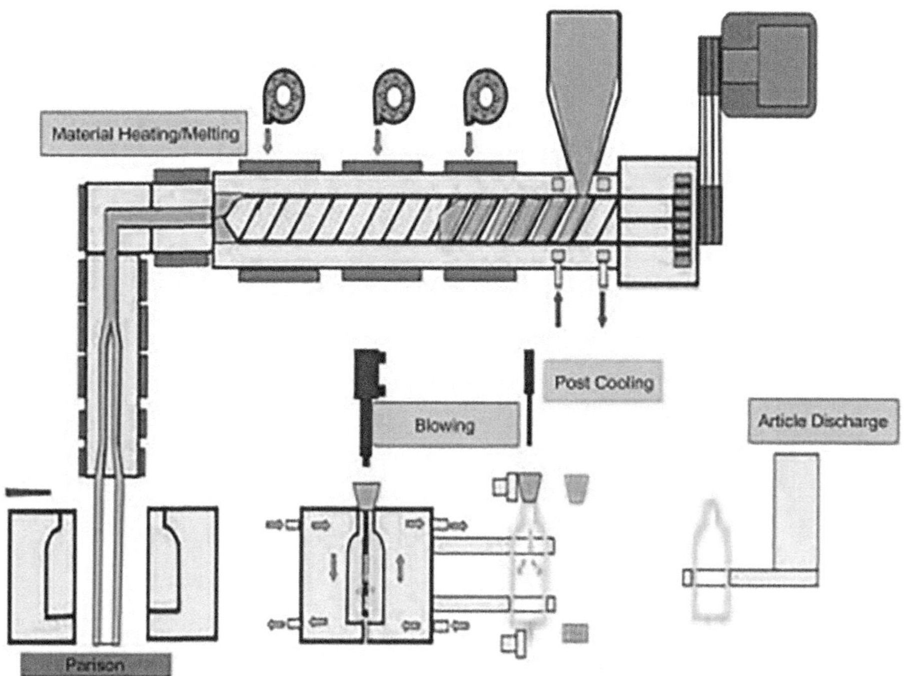

FIGURE 4.8   Different stages of the extrusion-blow molding process.

(Adapted from [7].)

lower initial investment costs for machinery and molds compared to other methods like injection molding and extrusion-blow molding. The most commonly used polymer in rotational molding is polyethylene (PE) (90% of applications); polypropylene (PP), polycarbonate (PC), polyamide (PA), and polyvinyl chloride (PVC) are also utilized. To be suitable for rotational molding, PE must be in powder form with particle sizes ranging from 100 to 500 μm.

The process involves rotating a heated mold within an oven along two perpendicular axes. During rotation, every point on the mold's internal surface traverses the entire volume, intermittently contacting the plastic powder. The rotational molding cycle comprises four stages (Figure 4.9). It begins with material loading, where a measured quantity of plastic powder (typically polyethylene) is loaded into a mold, often made of aluminum or steel to ensure good thermal conductivity. Next, during heating and fusion, the mold is tightly sealed and placed in an oven, rotating slowly along two axes. Continuous rotation enables the plastic powder to melt and evenly coat the mold's inner walls. Cooling follows as the melted plastic forms a uniform layer inside the mold and is transferred to a cooling station, potentially hastened by fans or water spray. During this phase, the plastic solidifies, assuming the mold's shape. Finally, in demolding, once the plastic has fully hardened, the mold is opened to extract the finished piece. The product undergoes inspection, and any excess material is trimmed off.

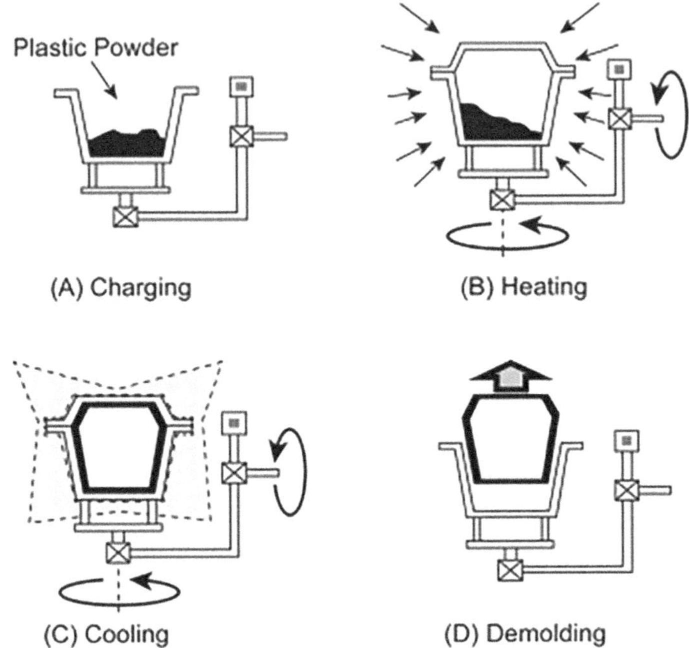

FIGURE 4.9   Biaxial rotomolding machine. (A) Charging; (B) Heating; (C) Cooling; (D) Demolding [8].

The technique described is utilized for processing thermoplastic polymers, offering both advantages and challenges. Rotational molding presents several distinct advantages over traditional molding processes. First, the molds used are simple and inexpensive because the process does not involve high pressures, eliminating the need for high-strength molds. Additionally, since no applied pressure is involved, the parts produced do not exhibit residual stresses from mechanical origins. This process also minimizes material waste, making rotational molding not only straightforward but also cost-effective; a rotational molding machine costs up to three times less than an equivalent injection molding machine. Complex-shaped articles, including large, thick pieces and technical parts like multilayered or foam-filled components, can be efficiently produced via rotational molding. The relatively long cycle time allows for concurrent chemical modifications, leading to innovations such as reactive rotational molding. It's also easy to incorporate plastic or metal inserts into the mold, and seam lines from mold parting planes can be avoided. Moreover, an excellent surface finish can be achieved with this method. However, rotational molding also poses significant challenges. The cycle time is considerably longer than other processes, resulting in low production throughput. Only a limited range of materials are suitable, with approximately 90% of rotational molded parts made from polyethylene, although the diversity of usable materials is increasing. Additionally, materials must be ground into powder form, increasing raw material costs, and for those with very high processing temperatures, the risk of degradation is not negligible.

Reactive rotational molding, or chemical rotational molding, is an advanced variation of the traditional rotational molding process that not only forms plastic parts but also facilitates chemical reactions during the molding process. This technique is particularly suitable for manufacturing composite products or incorporating specific properties into finished products. The basic principle of reactive rotational molding is similar to traditional rotational molding, where powdered polymer is placed in a mold that is then heated and rotated. The major difference is that in reactive rotational molding, chemical precursors are also included in the mold. These precursors react under heat, forming polymers through polyaddition or polycondensation reactions. During mold heating and rotation, the chemical reaction occurs concurrently with polymer melting. This may involve forming new chemical bonds or polymerizing monomers to create more complex structures. The mold is then cooled to solidify the material and halt the reaction, after which the finished piece is extracted from the mold.

Reactive rotational molding offers numerous advantages, including customized material property adjustments. It allows for integrating specific characteristics such as UV resistance, chemical resistance, or enhanced flexibility by adjusting the reactants or additives used. This process also streamlines production by directly integrating the chemical reaction into the molding process, eliminating the need for posttreatments or additional manufacturing steps, thus reducing costs and simplifying production. Applications of reactive rotational molding are diverse, including manufacturing composite automotive parts, containers for chemicals, and structural elements requiring specific mechanical properties. Functionalized plastics resistant to high temperatures or with improved insulation properties are also suitable for this process. However, reactive rotational molding presents challenges such as controlling the chemical reaction and requiring precise management of temperature and reaction time to ensure the desired quality and properties of the finished product. Additionally, the use of special monomers or precursors can increase raw material costs compared to standard rotational molding. In summary, while reactive rotational molding significantly expands the capabilities of traditional rotational molding by enabling advanced customization of material properties, it requires meticulous process control to optimize outcomes.

4.10.2.3 Composite Layer

The Type IV tank consists of an advanced composite shell with a polymer matrix reinforced by carbon or glass fibers, crucial for internal pressure resistance and overall structural strength. This composite layer, wrapped around the liner in specific orientations, maximizes the durability and strength of the tank, playing a crucial role under high pressures. In the manufacturing of the outer layer of the compressed hydrogen tank, epoxy resin or unsaturated polyester is typically used, which, once cured, forms a three-dimensional structure.

4.10.2.3.1 Polyepoxy Matrix

Epoxy resin, or simply epoxy, is a thermosetting polymer widely used in numerous industrial and commercial fields due to its exceptional properties. The curing process of epoxy resin involves the formation of a rigid three-dimensional network, imparting a wide range of desirable characteristics. Epoxy resins are synthesized by cross-linking an epoxy oligomer with a hardener. The most commonly used epoxy oligomer is diglycidyl ether of bisphenol A (DGEBA), which contains epoxy groups (oxirane) capable of reacting with various hardeners to form a three-dimensional polymer network. Amine-based hardeners are commonly used to create covalent bonds, resulting in a dense and rigid material with high mechanical properties and good temperature resistance. Alternatively, anhydrides used as hardeners produce materials with excellent chemical resistance and are often chosen for applications requiring strong corrosion resistance.

Epoxy resins are distinguished by their exceptional characteristics once cured, forming a strong and rigid material with high mechanical strength, making them ideal for structural and load-bearing applications. They effectively adhere to a wide variety of substrates, including metals, glass, and most plastics, making them perfect for adhesives and coatings. Their three-dimensional network provides an excellent barrier against chemicals and moisture, essential for protective coatings and applications in harsh environments. Epoxy is also a preferred electrical insulator, suitable for electrical and electronic equipment. Its dimensional stability and minimal shrinkage during curing allow for precise dimensional tolerances in molded parts. Moreover, epoxy composites can withstand relatively high temperatures, although performance may vary depending on the curing system used. In summary, due to its robust mechanical properties, chemical resistance, and versatility, epoxy is a preferred material for a multitude of applications, ranging from structural composites to protective coatings, industrial adhesives, and electrical applications.

4.10.2.3.2 *Unsaturated Polyester Matrix*

Polyester resin is a thermosetting polymer that forms a three-dimensional network during curing. Its synthesis involves a cross-linking reaction primarily between two components: unsaturated polyester resin and styrene. The resin, derived from the reaction of a polyunsaturated acid like maleic acid with a diol such as glycol, forms the main chain of the polymer and contains carbon-carbon double bonds crucial for curing. Styrene acts as the cross-linking monomer, mixing with the resin to lower viscosity and facilitate processing. During curing, initiated by a catalyst like peroxide, the double bonds in both styrene and the resin react to create a rigid and solid polymer network.

Unsaturated polyester resin is characterized by its ease of processing, where resin viscosity can be adjusted by adding styrene, enabling techniques such as contact molding, pultrusion, or injection molding. Once cured, this material offers good chemical resistance against many solvents, making it ideal for applications like chemical storage tanks and pipelines. Although its mechanical properties and rigidity are generally lower than those of epoxy composites, unsaturated polyester resin withstands weathering well, making it suitable for outdoor use in components such as boat parts and vehicle bodywork. Moreover, being less expensive than other thermosetting resins like epoxies, it is attractive for a wide range of industrial and construction applications. Its versatility and cost-effectiveness also promote its use in manufacturing reinforced fiberglass products for diverse applications, from household items to complex structures like boats and vehicles. Thus, unsaturated polyester resin is a popular choice in many industrial sectors due to its properties suited for various environments and uses.

Unsaturated polyester is often more affordable than epoxy, which can be crucial for large-scale applications with limited budgets. In terms of mechanical strength, epoxy demonstrates higher values, indicating a better ability to withstand mechanical stresses, making it preferable for demanding structural applications. Chemical resistance is also a strong point of epoxy, essential for storage applications exposed to various chemicals, including hydrogen. Epoxy's low permeability is vital for minimizing hydrogen leaks and maintaining energy efficiency, while its ability to withstand repeated pressure cycles without compromising structural integrity ensures the safety and longevity of hydrogen tanks. Moreover, epoxy stands out for its high-temperature stability, which is also a major asset, reducing the risks of failure under heat. These properties make epoxy the material of choice for critical applications such as compressed hydrogen tanks, where its superiority in mechanical properties, chemical resistance, and durability ensures safety and efficiency.

4.10.2.3.3 *Thermoplastic Matrix*

In the design of compressed hydrogen tanks, the use of thermoplastic matrix composites offers several significant advantages over traditionally used thermoset matrices. Thermoplastics are particularly valued for their recyclability; unlike thermosets, they can be melted and reshaped, facilitating material recycling. Additionally, they exhibit better impact resistance and greater fatigue durability, crucial properties for applications requiring high resilience. Another notable advantage of thermoplastics is their reduced processing time as they require less time to harden, which can significantly decrease production cycle times.

However, integrating long fibers with thermoplastic matrices, which have high viscosity when molten, presents specific challenges. Ensuring a homogeneous mixture of long fibers with the molten polymer is difficult due to this high viscosity, potentially compromising the mechanical properties of the final composite. Moreover, controlling the distribution and orientation of fibers in the matrix is crucial to maximize mechanical properties, and the high viscosity makes this control more challenging. Furthermore, the high temperatures and pressures required to process thermoplastics can damage fibers or affect their performance, adding complexity to the manufacturing process. These challenges require careful attention during the design and manufacturing of thermoplastic matrix composites to ensure their effectiveness and reliability in critical applications such as compressed hydrogen tanks.

In the field of composites used for compressed hydrogen tanks, various thermoplastic polymers are favored for their specific properties that meet the rigorous requirements of these applications. HDPE is particularly valued for its chemical resistance and ease of processing, while PA is appreciated for its wear resistance and excellent mechanical properties. Polyetheretherketone (PEEK), although more expensive, excels in thermal and mechanical resistance, making it ideal for applications where performance under high pressure is critical.

For effective integration of long fibers into these thermoplastic matrices, several specialized techniques are implemented. For example, filament winding involves wrapping fibers around a mold, with thermoplastics applied as film or powder that melts and impregnates the fibers under heat. Resin Transfer Molding (RTM) has been adapted for thermoplastics by introducing rapid heating systems to manage the high viscosity of these polymers. Pultrusion, on the other hand, has been modified to maintain polymers at an adequate temperature during the process through adjustments in heating systems.

These integration methods are crucial as they overcome challenges related to the high viscosity of thermoplastics and ensure optimal exploitation of fiber properties in the final composite. Ongoing innovation in this field is critical to improving processing techniques and the final properties of composite materials, thereby increasing the viability of these technologies for critical applications such as hydrogen storage. Currently, the use of thermoplastic matrix composites for compressed hydrogen tanks is in the R&D phase. While theoretical advantages and some initial results are promising, there is still much work to be done to overcome technical challenges and fully assess the commercial and technical viability of this approach. This R&D effort is essential to ensure that the developed solutions are not only effective but also safe and economically viable on an industrial scale.

4.10.2.3.4 Carbon Fiber Reinforcement

Carbon fibers are highly prized advanced materials known for their strength and lightness, particularly in applications such as hydrogen tanks. These fibers are manufactured from organic polymer precursors such as polyacrylonitrile (PAN), pitch, or rayon. The manufacturing process involves several critical stages. First is oxidation, where the precursor is heated in air at approximately 200–300°C to stabilize its structure. Next, is carbonization, during which the stabilized material is heated at high temperatures, up to 1500°C in the absence of oxygen, to remove noncarbon elements and produce a nearly pure carbon structure. Some applications require graphitization, where the fibers are heated to 3000°C to refine the crystalline structure. Surface treatment may also be applied to enhance adhesion with polymer matrices in composites, followed by winding the final fibers onto spools.

There are primarily two types of carbon fibers: high modulus fibers, optimized for rigidity and often used in aerospace industries, and high strength fibers, designed to maximize tensile strength and used in applications requiring high mechanical strength. Carbon fibers are extremely thin, typically with a diameter of 5 to 10 microns. They can be used individually or assembled into yarns or fabrics, depending on the specific application requirements. Carbon fibers possess exceptional properties that make them superior to many traditional materials. They offer high tensile strength, surpassing steel at equal weight, while having a density approximately four times lower than that of steel, making them extremely lightweight. Additionally, their excellent fatigue and corrosion resistance make them ideal for hostile environments. Carbon fibers also exhibit good thermal and electrical conductivity properties, suitable for specific applications.

Due to these unique properties, carbon fibers are widely used across various sectors. In aerospace, they are employed in aircraft and satellite components. In the automotive sector, they are used for chassis parts and sports car bodies. Sports benefit from these materials for bicycle frames, tennis rackets, and fishing rods. In construction, they reinforce concrete structures and bridges, while in the energy sector, they are used for wind turbine blades. Specifically in hydrogen storage tanks, carbon fibers play a crucial role by forming composite layers that must withstand extremely high pressures while remaining lightweight. Their ability to withstand repeated cycles of pressurization and depressurization without significant fatigue is essential for ensuring the safety and efficiency of hydrogen storage. These characteristics make carbon fibers indispensable for the future of energy storage technologies and the development of advanced materials.

4.10.2.3.5 Glass Fiber Reinforcement

In the context of hydrogen storage tank manufacturing, long glass fibers, also known as fiberglass, are commonly utilized across various industries due to their moderate cost, strength, and flexible fabrication capabilities. The manufacturing process of fiberglass, known as spinning, includes several key stages. Initially, raw materials such as silica sand, limestone, and sodium carbonate are melted at high temperatures to achieve a homogeneous liquid. This molten glass is then extruded through fine nozzles to create individual fibers. These fibers are rapidly cooled to prevent crystallization and undergo surface treatment to enhance their adhesion to resins and other polymers used in composites. There are different types of fiberglass, primarily classified based on their chemical composition and specific properties. E-glass fibers, used mainly in electrical applications, are corrosion-resistant. S-glass fibers, known for their high mechanical strength, are often employed in demanding structural applications. C-glass fibers offer high chemical resistance, making them suitable for corrosive environments.

The dimensions of fiberglass vary, with diameters ranging from a few microns to several tens of microns, depending on the application. These fibers can be used individually or assembled into yarns, fabrics, or mats, offering a wide range of potential applications. Fiberglass possesses a range of properties that make it valuable in diverse industrial applications. They provide good tensile strength, although lower than carbon fibers, and are lighter than metals, which helps reduce structure weight. Additionally, they exhibit good thermal stability up to moderately high temperatures and excel as electrical insulators. Their corrosion resistance also makes them ideal for harsh environments. These fibers find utility in numerous sectors, including automotive and transportation, where they are used for body components and internal parts. In building and construction, they are used for insulation panels and reinforced concrete structures. The electrical and electronic sector also benefits from their properties for manufacturing printed circuits and other insulators. In the sports and leisure industry, they are employed to manufacture sporting equipment such as skis, boats, and fishing rods.

In the fabrication of hydrogen storage tanks, fiberglass is particularly advantageous for creating composite layers that withstand high pressures while remaining cost-effective. Their combination of mechanical strength, lightweight, and corrosion resistance allows for the production of efficient and durable tanks, which are more affordable compared to those made solely from carbon fibers. Thus, fiberglass plays a crucial role in many modern industries, significantly contributing to the safety and efficiency of energy storage technologies, especially in hydrogen storage systems. In Table 4.6, one type of carbon fiber (T700S) is compared to one type of fiberglass

TABLE 4.6

Summary of Properties of Major High-strength Fibers

Properties	Inorganic Fibers	
	T700S	Hiper-tex
Nature	Carbon	Glass
Stress at break (tensile); MPa	4,900	4,500[a]
Specific stress (tensile); $MPa.M^3.kg^{-1}$	2.7	1.8
Young modulus (GPa)	230	86[a]
Strain at break (%)	2.1	5.5[a]
Density $(g.cm^{-3})$	1.8	2.5[a]
Water absorption (%)		1
Degradation temperature (°C)	400 (with O_2)3 000 (without O_2)	1 200[a]

[a] Values estimated from information given orally or from neighboring fibers.

(Hiper-tex). We can observe the superiority of carbon fiber over fiberglass in terms of mechanical properties, especially regarding strength and stiffness, which are crucial elements for a high-pressure hydrogen tank.

The composites used in Type IV tanks offer several crucial advantages. First, they provide exceptional pressure resistance, essential for storing gases under high pressure. Additionally, these materials are significantly lighter than traditional metals, greatly facilitating the transport and installation of tanks. Moreover, durability is enhanced by the use of carbon and glass fibers, which effectively resist corrosion, fatigue, and impacts.

However, these advantages come with significant challenges. The cost of carbon fibers can significantly increase the total manufacturing cost of the tank. Furthermore, the sensitivity of some composite matrices to UV and moisture requires the application of protective coatings or the use of special formulations to prevent degradation. The manufacturing process itself is complex, requiring specialized equipment and technical expertize to ensure precise fiber orientation and effective curing. Manufacturing costs vary widely, influenced by the type of fiber, the resin used, and specific fabrication techniques such as filament winding and autoclaving, which require substantial investments in equipment and staff training. Although the composite layer of Type IV tanks is crucial for their performance and safety, it poses challenges in terms of cost and manufacturing complexity. Nevertheless, ongoing research and innovations in composite materials and their manufacturing processes could help overcome these obstacles in the future.

4.10.2.4 Composite Manufacturing Methods

The outer layer of Type IV tanks for compressed hydrogen, typically composed of composite materials, plays a crucial role in protecting and maintaining the structural integrity of the tank under high pressure. The fabrication method of this composite layer heavily depends on the geometry and shape of the tank. For cylindrical tanks, which are common in many applications, particularly in the automotive sector, two main techniques are used: filament winding and braiding. Filament winding is a method where fibers, often carbon or glass, impregnated with resin are wound under tension around a mandrel that is later removed after resin curing. This technique allows precise control over fiber orientation, which is essential for optimizing the tank's resistance to high internal pressures and external forces. Winding can be performed axially, helically, or in a combination of both, depending on specific requirements for strength and tank performance. Braiding, on the other hand, involves interlacing fibers to form a mesh structure that is then coated or impregnated with resin. This method can provide good uniformity and mechanical strength in all directions, which is beneficial for tanks that need to withstand multidirectional loads. Braiding is particularly suitable for complex shapes or when a homogeneous distribution of strength is needed around the entire tank.

These manufacturing methods are chosen not only for their technical efficiency but also for their economic viability, enabling the mass production of lightweight yet extremely strong tanks, essential for the safe storage of hydrogen under high pressure.

4.10.2.4.1   Filament Winding

Filament winding is an advanced manufacturing technique used to produce the composite layer of compressed hydrogen storage tanks. This method is particularly valued for its ability to create tanks that are lightweight, strong, and capable of withstanding very high pressures. It involves winding continuous fibers impregnated with resin around a mandrel, typically cylindrical in shape, which defines the internal form of the tank. The fibers used are typically carbon or glass fibers, known for their high strength and lightweight.

The process for manufacturing Type IV tanks involves several key steps designed to ensure the tanks can safely contain compressed hydrogen.

Preparation begins with the liner, which acts as the mandrel and internal barrier for hydrogen. This liner is typically made of plastic or another polymer material capable of holding hydrogen under high pressure. Impregnation follows, where fibers, usually carbon or glass, are impregnated with resin before winding. Impregnation can be done either through a bath process where fibers are immersed in resin, or through an inline process where resin is applied just before winding. Winding is the next step, where impregnated fibers are wound around the liner in a predetermined pattern. This may include helical and circular windings to strengthen the tank against internal and external pressures. The angle and tension of winding are precisely controlled to ensure optimal coverage and strength. After winding, the composite needs to cure. This curing process may occur in an oven where the tank is heated to a specific temperature to allow the resin to harden and form a solid composite. Finally, after curing, the tank undergoes finishing steps. These may include sanding the outer surface to prepare for additional coatings or conducting inspections to ensure structural integrity is maintained throughout the manufacturing process.

The filament winding process used to manufacture compressed hydrogen storage tanks is influenced by several key parameters that must be carefully controlled to ensure the integrity and performance of the finished product. Fiber tension during winding is crucial, requiring a balance to maintain alignment and contact with the liner without damaging it, typically made of flexible polymer. The winding angle varies to meet specific strength needs: closer angles to the longitudinal axis enhance resistance to longitudinal stresses, while 90-degree angles bolster resistance to radial stress. Resin-to-fiber ratios affect composite mechanical properties, ensuring adequate resin for cohesion without brittleness. Precise curing temperature and duration are crucial for achieving desired material properties, impacting strength, chemical resistance, and dimensional stability. Mastering these process parameters during filament winding is critical for producing hydrogen storage tanks meeting safety and performance standards and maximizing system lifespan and efficiency.

In the filament winding technique for manufacturing compressed hydrogen storage tanks, two main methods are used to apply fibers to the mandrel: dry winding and wet winding. These methods differ in when and how the fibers are impregnated with resin before being wound around the mandrel.

In dry winding, fibers are wound around the mandrel without resin. Resin impregnation occurs after the fibers are wound onto the mandrel. Dry fibers are wound according to a defined pattern and orientation. After winding, the resin is applied to the fibers already positioned on the mandrel. This can be done through spraying, dipping, or pressure injection into the wound material. Curing then takes place in an oven where the applied resin hardens to form the final composite material. This method allows better control over resin distribution uniformity, reducing the risk of trapped air between fibers, which can enhance composite quality. Challenges associated with dry filament winding, especially when resin impregnation occurs after winding, include the potentially laborious process requiring sophisticated equipment for effective impregnation. Additionally, managing resin viscosity, crucial for ensuring adequate and uniform penetration into the fiber fabric, can be complex. These factors require careful attention to optimize the quality and integrity of the finished tank.

In wet winding, fibers are impregnated with resin before being wound onto the mandrel. Fibers pass through a resin bath or device that coats them just before they are wound onto the mandrel. Resin-impregnated fibers are directly wound onto the mandrel in the desired orientation. The composite is then cured, often through heating, allowing the resin to solidify and bind the fibers together. The advantage of combining impregnation and winding in a single step includes simplifying the production process, potentially making manufacturing faster and more efficient. However, this method presents challenges such as the risk of variability in the resin-to-fiber ratio if impregnation control is not precise enough. Moreover, there is a potential for air inclusions in the composite material if impregnation is not perfectly homogeneous, which can affect the quality and performance of the final product (Figure 4.10).

The choice between dry and wet filament winding depends largely on the product specifications, quality requirements, and existing production infrastructure, with each method offering its own advantages and challenges. For the manufacture of compressed hydrogen tanks, the wet winding method is preferred due to several significant advantages. It allows precise and immediate control of the resin-to-fiber ratio during impregnation, crucial for achieving optimal mechanical properties of the composite material, and ensures consistency in composite quality by reducing risks of defects such as air pockets or dry areas that could compromise the tank's strength

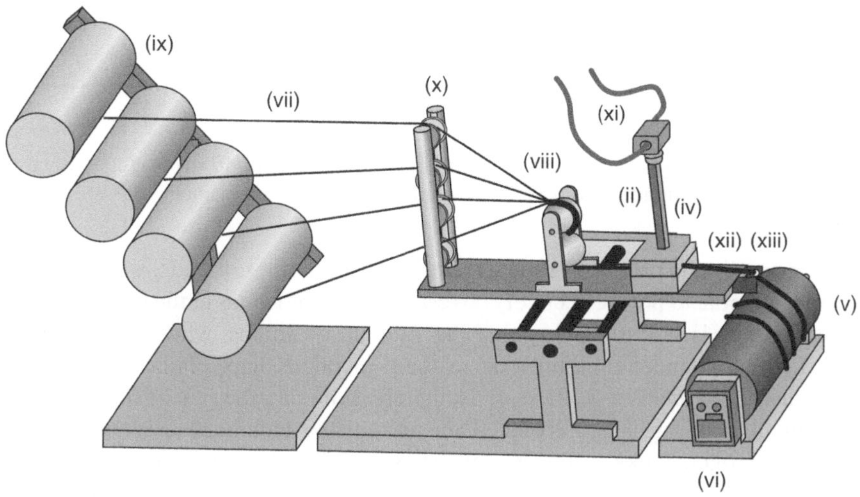

FIGURE 4.10 Wet filament winding.

(Adapted from [9].)

and integrity under high pressure. Furthermore, combining impregnation and winding in a single step simplifies the manufacturing process, reduces cycle times, and enhances production efficiency, which is particularly beneficial for large-scale production by lowering costs and improving the ability to meet high demand. Pre-impregnation of fibers ensures uniform resin distribution, essential for all fibers to effectively contribute to the tank's mechanical strength, while minimizing irregularities that may occur with post-winding impregnation. Finally, applying resin before winding minimizes the risk of air inclusion and bubble formation, critical defects to avoid in tanks required to withstand very high pressures without leakage, while ensuring fibers are fully saturated with resin, thus avoiding dry zones and maintaining full coverage.

Furthermore, filament winding methods play a crucial role in tailoring compressed hydrogen storage tanks to meet specific requirements of strength and pressure. Each winding technique is carefully designed to reinforce the tank to withstand various types of stresses, and they are often used together to maximize tank performance under high pressure. Helical winding as shown in Figure 4.11, involves winding fibers in a spiral around the mandrel at an angle that is neither parallel nor perpendicular to the mandrel axis, forming a helical structure that diagonally wraps around the tank. This method is particularly effective in fortifying the tank against longitudinal stresses along the tank axis, as well as circumferential stresses around its circumference. It is commonly used for the main bodies of cylindrical tanks. Circular winding, also known as "hooping," entails winding fibers strictly perpendicular to the mandrel axis, creating rings of fibers that encircle the tank. This type of winding is essential for increasing the tank's resistance to radial internal pressures, which push outward from the center of the tank, and plays a crucial role in preventing tank rupture under high pressure.

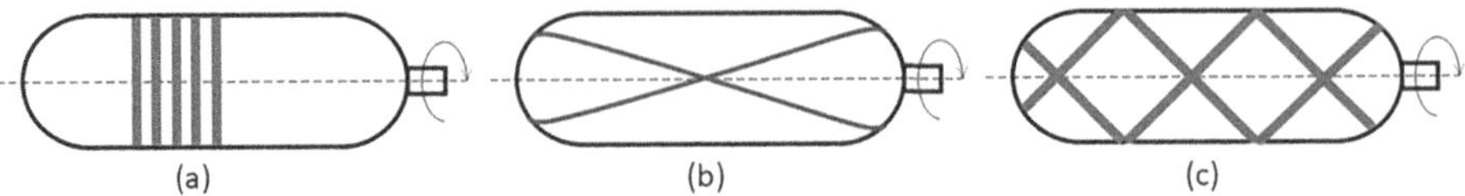

FIGURE 4.11 Schematic description of winding patterns and uniform coverage: (a) circular winding, (b) polar winding, and (c) helical winding pattern of the cylinder.

Polar winding is primarily used to reinforce tank domes or ends, where fibers are wound from pole to pole, effectively covering the curved areas at the ends of the tank. This technique is vital for enhancing the strength of domes, which are particularly prone to failure under pressure due to their rounded shape. Polar winding helps evenly distribute pressure at these ends, thereby reducing the risk of rupture.

Filament winding is an advanced manufacturing technique essential for compressed hydrogen tanks, offering numerous benefits despite inherent challenges. This method is particularly valued for its ability to produce tanks that withstand very high pressures, crucial for safe hydrogen storage. In addition to strength, lightweight construction is another major advantage, as composite materials used in filament winding significantly reduce weight compared to traditional metals. The technique also offers remarkable design flexibility, enabling the creation of tanks in various geometries and sizes, essential for adapting to diverse applications and space requirements.

However, filament winding presents certain challenges. The cost of materials, especially carbon fibers, is high, which can significantly increase the production cost of tanks. Moreover, the technique demands high technical precision during manufacturing, requiring rigorous control to maintain quality and performance. While the method is flexible, it can be complex to apply for noncylindrical shapes, sometimes limiting its application. In terms of costs, although filament winding is expensive due to the use of composite materials and the need for specialized technologies, the savings in weight and the enhancement in performance often justify the initial high investment. The benefits in terms of durability, energy efficiency, and vehicle safety can largely offset these initial costs. Despite challenges related to costs and technical complexity, filament winding remains a preferred method for manufacturing compressed hydrogen tanks. Its advantages in strength, lightweight construction, and safety make it an indispensable solution for applications requiring high performance.

4.10.2.4.2 Braiding

Braiding of composite materials is an advanced manufacturing technique that reinforces structures by intertwining continuous fibers in specific patterns. This process involves using fibers such as carbon, glass, or aramid, which are braided around a core or mold to create a predefined geometry. Not only does this method enhance the mechanical strength and durability of structures but it also reduces their weight compared to traditional materials. The technique encompasses various principles, stages, and technologies, offering numerous advantages as well as challenges, particularly in the development of compressed hydrogen tanks, where these qualities are crucial.

To understand in detail the stages of braiding composite materials, it is important to consider each phase of manufacturing, as each step directly influences the quality and performance of the final product. Computer-Aided Design (CAD) software is essential for modeling the geometry of parts, enabling precise designs and experimentation with different configurations to tailor the shape to specific application constraints. Before proceeding to physical manufacturing, finite element simulations are commonly used to anticipate how the part will react under various loads. This crucial step helps identify potential design weaknesses and allows for adjustments to enhance the performance and durability of the final product. Fiber selection is fundamental in composite material manufacturing, with each type—whether carbon, glass, or aramid—possessing unique properties that impact the strength, elasticity, and weight of the final product. Fiber preparation is equally critical; fibers often need treatment through coating or impregnation with a binding agent to facilitate the braiding process and ensure optimal adhesion to the resin. This preparation is essential to maintain the integrity and quality of the finished composite material. The mold or mandrel is the tool around which fibers will be braided, defining the shape of the finished piece. Precision in mold fabrication is critical, as any inaccuracies will affect the final product. Using a braiding machine, fibers are interlaced around the mold. This machine enables meticulous control over fiber orientation and tension, crucial for ensuring structural uniformity and optimizing mechanical properties. After braiding, resin is applied to bind the fibers together. The piece then undergoes thermal treatment or pressure (or both) to cure the resin, solidifying the entire structure and efficiently transferring mechanical loads throughout the composite. Demolding involves removing the piece from the mold after the resin has cured, marking a key step in the manufacturing process. Once demolded, the piece may require additional adjustments, such as trimming edges to achieve precise dimensions. Surface finishes are also often applied not only to enhance aesthetic appearance but also to reinforce specific physical properties like wear or corrosion resistance. These final steps are crucial to ensuring the finished product meets quality standards and required technical specifications. Each stage is designed to maximize the quality and efficiency of the final product by leveraging the advantages of composite materials while minimizing inherent manufacturing defects.

Technical aspects play a pivotal role in the advanced techniques of braiding. Modern braiding machines (Figure 4.12) are adept at handling multiple fiber yarns simultaneously, allowing for the creation of intricate structures with multi-axis capabilities. For instance, 3D braiding machines can produce composite preforms with fibers oriented in three dimensions, significantly enhancing the mechanical properties of the final product. Prior to braiding, meticulous design and modeling are essential to determine the precise fiber orientation and quantity required to meet stringent performance criteria. This process often involves the use of CAD and Finite Element Analysis (FEA) software to optimize the structural integrity before manufacturing begins. Specific braiding techniques like axial braiding, which aligns fibers along the part's longitudinal axis, are ideal for components subjected to tensile stresses. Conversely, radial braiding involves fibers braided in a circular pattern around the part, commonly applied in the fabrication of tubes and shells, highlighting the versatility and tailored application of braiding technologies in composite manufacturing.

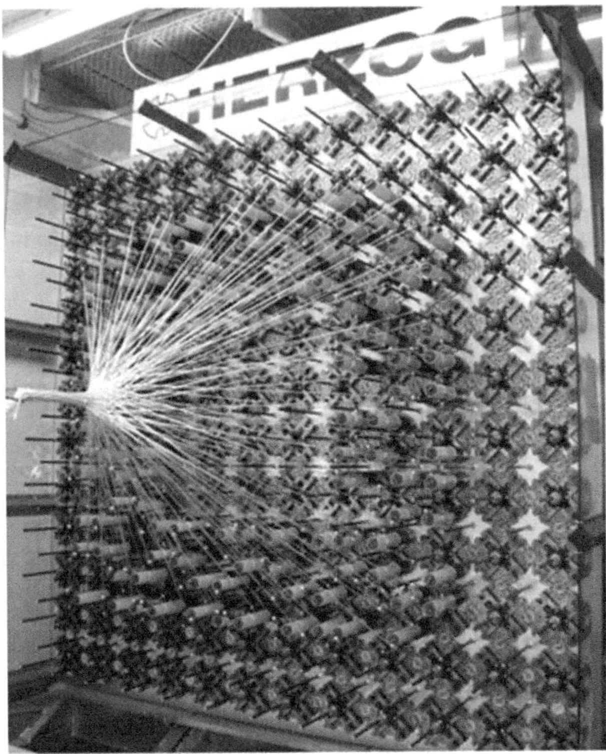

FIGURE 4.12   9-Module Herzog 3D rotary braiding machine for the manufacture of composite material parts.

(Adapted from ref. [10].)

Understanding the fabrication of braided composite materials involves analyzing key parameters that influence their performance. Fiber type selection—such as carbon for strength and lightness, glass for cost-effective strength, and aramid for impact resistance and flexibility—tailors the composite's suitability for diverse applications. Fiber orientation directs mechanical properties, with longitudinal alignment enhancing tensile strength and radial braiding bolstering compression and shear resistance. Increasing braiding density strengthens components but may add weight, necessitating a balance between robustness and lightweight design. Resin choice, like epoxy for strong bonding and thermal resistance, complements fibers to optimize composite durability and chemical resilience in harsh conditions, crucial for applications like chemical storage tanks.

Braiding technologies in composite material manufacturing encompass 2D and 3D techniques along with robotics, each serving specific structural and design requirements. 2D braiding is ideal for creating flat or tubular shapes, interlacing fibers primarily on a horizontal plane suitable for applications needing strength in a single direction. In contrast, 3D braiding expands on this by integrating fibers across multiple axes, enhancing strength in all dimensions crucial for unpredictable loads in the aerospace and automotive sectors. Robotics further revolutionizes braiding by enabling precise fiber manipulation around complex shapes, optimizing strength and flexibility for custom components like medical implants or aerospace parts requiring seamless integration into existing structures. These technologies collectively advance composite manufacturing capabilities, catering to diverse industrial needs with enhanced performance and design flexibility.

Woven composite materials offer several significant advantages that make them particularly appealing for a variety of industrial applications. First, they combine lightweight with high strength, crucial for sectors like aerospace and automotive where weight reduction can enhance energy efficiency and performance. Additionally, these materials offer great design flexibility, allowing engineers to create complex structures tailored to specific needs. Finally, they possess excellent energy absorption properties and load distribution, making them ideal for applications requiring increased resilience against impacts and dynamic loads. However, the use of woven composite materials also presents notable challenges. The cost of fibers and braiding technology can be high, making these materials less accessible for some projects or large-scale productions. Design complexity for nonstandard shapes can also be problematic, requiring advanced CAD software and specific skills in simulation and modeling. Moreover, manufacturing woven composites demands specialized expertize in both design and production processes, limiting the availability of skilled labor and increasing training and development costs. In the R&D domain, the focus is on optimizing the manufacturing processes of woven composite materials, reducing costs, and enhancing their properties, particularly for critical applications in the aerospace and energy sectors. These efforts aim to maximize production efficiency while improving material performance to meet rigorous technical requirements in these industries.

Compressed hydrogen tanks represent a particularly promising area for the use of woven composite materials. These tanks require exceptional strength to withstand high pressures yet must remain lightweight to be effectively integrated into vehicles. Through weaving techniques, it is possible to manufacture cylindrical tanks with fibers strategically aligned to optimize resistance to internal pressure and minimize the risk of rupture. This approach not only enhances safety but also contributes to improved energy efficiency of vehicles.

4.10.2.5   Metallic Insert

Type IV compressed hydrogen tanks are engineered to store hydrogen under extremely high pressures, typically up to 700 bars or more. These tanks feature an inner liner made from plastic materials like polyethylene, surrounded by an outer shell composed of composite materials such as carbon fiber. At the heart of these tanks, the metal insert plays pivotal roles essential for both performance and safety. First, it facilitates the loading and unloading of hydrogen by providing a crucial channel for efficient passage within the tank, which is vital for effective hydrogen management. Second, the metal insert serves as a robust connection interface between the inner plastic liner and external metallic components, ensuring reliable connections for valves, regulators, and other critical system elements. Third, it maintains necessary seals to prevent hydrogen leaks, particularly at high-pressure and connection points, thus safeguarding operational safety and efficiency.

Surface treatments applied to the metal insert further enhance its performance and longevity. Techniques such as Physical Vapor Deposition (PVD) and CVD are utilized to improve surface properties, while anodizing is employed for aluminum alloys to increase corrosion resistance. Sandblasting prepares the surface for coatings by creating a rough texture that enhances adhesion, and surface oxidation forms a protective layer that enhances corrosion resistance and reduces chemical reactivity. The geometry of the metal insert is meticulously designed to optimize mechanical strength while minimizing overall weight. This design consideration is crucial to meet the specific requirements of each application and the tank's overall structural integrity. Complex shapes are tailored to efficiently integrate necessary interfaces and connection points, thereby maximizing structural robustness and minimizing potential weak spots. Materials used for the metal insert typically include stainless steel, aluminum alloys, and sometimes titanium alloys. These materials are chosen for their excellent corrosion resistance properties and lightweight nature, crucial factors where weight reduction and durability are paramount. Their compatibility with high-pressure hydrogen ensures minimal risks of chemical reactions or material degradation, enhancing the safety and reliability of Type IV compressed hydrogen tanks in various applications, from automotive to aerospace industries.

4.10.3 MAIN CHARACTERISTICS OF TYPE IV COMPRESSION HYDROGEN TANK

4.10.3.1 Hydrogen Impermeability

Type IV hydrogen tanks are at the forefront of clean energy storage technologies, using composite materials to store hydrogen under high pressure. These tanks play a crucial role in the viability of hydrogen vehicles, offering a lighter and stronger alternative compared to traditional metal tanks. In this context, understanding diffusion, solubility, and permeability phenomena is essential for optimizing their design and manufacturing processes.

Diffusion is the process by which gas molecules move from an area of high concentration to an area of low concentration. The diffusion of gas into a solid polymer is significantly influenced by various physicochemical characteristics of the material, as well as certain geometric aspects of the polymer piece being manufactured. Consider first the glass transition temperature (T_g) of the polymer, which is a crucial indicator of its physical state. Below T_g, the polymer is in a glassy and rigid state, limiting its diffusion capabilities. Above this temperature, polymer chains gain mobility, thereby increasing the available free volume for gas molecule movement, facilitating their diffusion through the material. Linked to this, free volume, which represents the spaces between molecules in a polymer, allows gas molecules to migrate more easily through the polymer when this volume is larger, typically under conditions where the polymer is above its T_g. The polymer's crystallinity also plays a crucial role in gas diffusion. Crystalline regions, which are denser and more compact, act as barriers to diffusion, whereas amorphous regions, with their less ordered structure, allow easier diffusion due to their larger free volume. Molecular weight distribution also influences diffusion. A polymer with a wide molecular weight distribution may exhibit a variety in molecular density, creating inconsistencies in free volume that can affect diffusion. Conversely, a polymer with a narrow molecular weight distribution tends to have more homogeneous properties, facilitating a more precise prediction of gas diffusion. The thickness of the polymer piece is an important geometric factor. A thicker piece provides a greater barrier to diffusion, increasing the distance gas molecules must travel, thereby slowing their passage through the material. This relationship is well described by Fick's law (see Section 3.2.1.6), which states that the diffusion flux is inversely proportional to the thickness of the polymer, directly affecting the rate at which gases can diffuse through the material. Each of these factors must be considered when selecting or modifying polymers for specific applications where gas diffusion management is essential, such as in food packaging, gas separation technologies, or components in energy storage systems. In hydrogen tanks, diffusion is a key factor influencing hydrogen leakage through the liner material.

Solubility is the measure of a gas's ability to dissolve in a given material at specific temperature and pressure conditions. It determines the amount of hydrogen that can be stored in the solid polymer before saturation is reached according to Equation (4.1):

$$C = S * P \tag{4.1}$$

where C is the concentration of dissolved gas, S is the solubility coefficient, and P is the pressure.

The dissolution of a gas into a solid polymer is initiated by the process of gas diffusion through the polymer; without diffusion, no dissolution can occur. Once gas molecules diffuse through the polymer, their solubility depends significantly on the physical affinity between gas molecules and polymer chains. This affinity is the primary factor determining gas dissolution in the polymer. Such interaction promotes the formation of secondary physical bonds, such as hydrogen bonds or Van der Waals interactions. These bonds are crucial as they stabilize gas molecules within the polymer network, thereby increasing the gas's solubility in the polymer. Regarding the effect of molecular mobility on solubility, it is essential to note that while increased mobility of polymer chains can facilitate the initial interaction between gas molecules and the polymer, it can also influence the stability of formed bonds. In polymers where chains are highly mobile, the secondary physical bonds formed may not always be stable over the long term, potentially leading to a degradation in solubility. This high mobility can allow gas molecules to escape more easily from the polymer, thus reducing the long-term effectiveness of gas dissolution. Therefore, chain mobility must be optimized to ensure a balance between high initial solubility and stable gas retention in practical applications.

Permeability in hydrogen tanks is critical for minimizing gas loss and maintaining required pressure. It measures how easily gas can move through a material, influenced by diffusion and solubility coefficients. Barrer's law equation, $(P = D * S)$, links these factors, detailing gas transport through polymers. This relationship is pivotal in optimizing material properties for Type IV hydrogen tanks, where reducing hydrogen loss is essential for storage efficiency and safety. The diffusion and permeability of gases through polymer materials are influenced by several key parameters. Material thickness (d) affects the time required for gas to traverse the material but not the diffusion coefficient (D), which remains constant. Permeability (P) inversely depends on thickness according to Fick's law. Temperature (T) significantly impacts diffusion and solubility (S): higher temperatures increase diffusion coefficients (D) due to enhanced molecular mobility, while solubility (S) may decrease depending on the gas–polymer interaction. Pressure (P) directly influences gas solubility following Henry's law, where solubility (S) is proportional to pressure (P). Moreover, pressure affects permeability (P) by altering the amount of gas that can dissolve in the polymer matrix. Understanding these relationships is crucial for optimizing the design and performance of polymer materials in various applications, including gas storage and transportation systems. As an example, Figure 4.13 shows the effect of pressure on the permeability coefficient of HDPE.

These interactions illustrate the complexity involved in the design and operation of Type IV hydrogen storage tanks. Optimizing these parameters according to specific application requirements is crucial to ensure the performance, efficiency, and safety of the hydrogen storage system.

4.10.3.2   Resistance to Very High Pressure

The Type IV tank, which is used in severe conditions (an operating pressure of 700 bars), must withstand a test pressure of 1575 bars without damage. To achieve this performance, studies have been conducted over several years including both theoretical and experimental approaches that are explained as follows.

4.10.3.3   Theoretical Approaches

Finite Element Modeling (FEM) is a powerful and widely used method for analyzing the mechanical behavior of Type IV compressed hydrogen tanks. These tanks are composed of complex composite layers, including a polymer matrix reinforced with carbon fibers. The FEM approach allows for the precise simulation of these

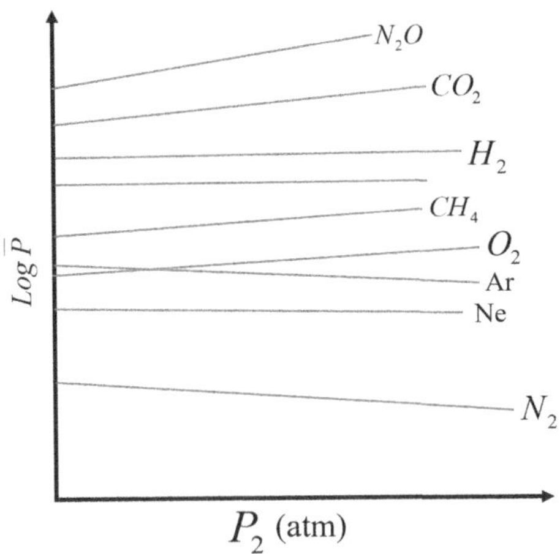

FIGURE 4.13   Influence of pressure on the permeability coefficient of HDPE to gases at room temperature.

structures' responses under various loading conditions, providing essential insights for their design and optimization. The use of simulation software such as ANSYS and Abaqus is at the heart of the FEM approach. These software tools enable the creation of detailed numerical models of the tanks, capturing the complex geometry and heterogeneous materials that comprise them. Users can define the material properties of the different layers, boundary conditions, and applied loads, such as internal pressure or external stresses. Thanks to these software's capabilities to solve complex mechanical equilibrium equations, it is possible to predict how the tank will react under different service conditions.

Modeling the composite layers is particularly crucial for Type IV tanks. Each composite material layer can have distinct mechanical properties, depending on the orientation of the fibers and the nature of the polymer matrix. In an FEM simulation, these layers are modeled individually and then assembled to form an overall composite structure. The interactions between the matrix and the reinforcing fibers are modeled to capture the load transfer mechanisms and deformation behaviors. The simulation software allows for specifying nonlinear behavior models for composite materials, accounting for phenomena such as plasticity, creep, and delamination. Stress and deformation analysis is a key step in understanding the tank's performance under various loads. Simulation results provide detailed maps of stresses and deformations throughout the tank's structure. This enables the identification of critical areas where stresses are maximal and where deformations might exceed safe service limits. By identifying these critical areas, engineers can predict potential failure modes and take measures to reinforce these areas or modify the tank's design to improve its reliability. FEM simulations can also be used to study the effects of different loading scenarios, including cyclic loads and accidental impacts, to ensure the tank can withstand real operational conditions.

The theory of elasticity and plasticity plays a vital role in studying the mechanical behavior of Type IV compressed hydrogen tanks. This theory is initially used to analyze the elastic deformations of the tank under pressure. It allows for describing the relationship between applied stresses and the resulting deformations for both the composite material components and the internal polymer liners. By applying equilibrium and compatibility equations and Hooke's laws to the various materials of the tank, the distribution of stresses and deformations under normal loading conditions can be determined. This ensures that the tank operates correctly within its elastic range, where deformations are reversible. When the applied stresses exceed the elastic limit of the materials, the theory of plasticity becomes necessary to model the behavior beyond this limit. The composite materials and polymers used in these tanks often exhibit complex behavior, involving phenomena such as plasticity, delamination, and rupture. Plasticity models, like those based on the von Mises or Tresca criteria, allow for predicting the response of materials under higher loads. These models incorporate flow criteria and evolution rules that describe how materials deform irreversibly.

Delamination, in particular, is a major concern in composite tanks because it can significantly degrade mechanical strength. Therefore, plasticity models must be coupled with delamination analyses to predict the long-term behavior of the tanks under repeated load cycles. Additionally, rupture phenomena are studied to understand the conditions under which the tank might undergo catastrophic failure. This includes using crack propagation models and failure mechanisms to anticipate potential weak points and design more robust tanks.

By combining the theory of elasticity for reversible deformations and plasticity models for irreversible deformations and failure mechanisms, engineers can gain a comprehensive understanding of the mechanical behavior of Type IV compressed hydrogen tanks. This integrated approach is essential for ensuring the safety and reliability of hydrogen storage systems, particularly in applications where tanks are subjected to severe operating conditions and high load cycles.

Analytical methods offer valuable approaches for solving simplified problems and obtaining quick and efficient estimates of stresses and deformations. One such analytical approach is to use shell models to represent cylindrical tanks. Type IV tanks can be modeled as thin shells due to their relatively thin walls compared to their radius. Shell models simplify the three-dimensional elasticity equations into two-dimensional equations, making them easier to solve.

The use of plate and shell theory is particularly suited for estimating stresses and deformations in these tanks. Shell theory, an extension of plate theory, accounts for surface curvature and is thus more appropriate for cylindrical structures. By applying this theory, differential equations describing the stress distribution in the pressurized tank can be derived. These equations consider the effects of geometry and specific boundary conditions of the tanks, allowing for predictions of mechanical behavior under various loading conditions.

For cylindrical tanks, the thin shell theory equations can be used to estimate circumferential (hoop) and longitudinal stresses. These stresses are crucial as they determine the resistance to internal pressure. The analytical solutions obtained from these models help understand how the composite materials of the tank distribute loads and react to internal pressures.

Moreover, these analytical methods are useful for quick checks during the preliminary design of the tanks. They help identify critical areas prone to high stress concentrations, guiding engineers in improving design and material selection. Although these methods are based on simplifications, they provide an essential starting point for more detailed analyses, such as those performed by the Finite Element Modeling (FEM), which can then refine predictions and confirm the results of analytical approaches.

Analytical methods, particularly the use of plate and shell theory, play a crucial role in studying the mechanical behavior of Type IV compressed hydrogen tanks. They provide powerful tools for quickly and efficiently estimating stresses and deformations while offering a fundamental understanding of the tanks' response mechanisms under pressure.

4.10.3.4 Damage and Failure Models

In studying the mechanical behavior of Type IV compressed hydrogen tanks, modeling damage and failure is essential to ensure the safety and durability of these tanks, which are often subjected to severe service conditions.

The main damage mechanisms include matrix cracking, fiber breakage, and delamination. Matrix cracking occurs when the polymer resin binding the reinforcing fibers cracks under excessive stress. This type of damage can reduce the material's ability to transfer loads between fibers, thereby weakening the composite structure. Fiber breakage occurs when the reinforcing fibers, typically carbon or glass, break under high tensile stresses. This mechanism is often critical because the fibers play a primary role in the composite's strength. Finally, delamination is the separation of composite material layers, which can severely compromise the tank's structural integrity. This type of damage is particularly concerning in Type IV tanks.

Various failure criteria are used to predict and analyze these damage mechanisms. The Tsai–Wu criterion is one of the most commonly used criteria for composite materials. It combines the effects of normal and shear stresses to provide a failure surface in the stress space. This criterion helps determine if a given point in the composite material has reached the failure state, considering the complex interactions between the different stress components.

The Puck criterion, specifically designed for fiber-reinforced composites, distinguishes between fiber rupture and matrix rupture, providing a more detailed analysis of the different possible failure modes in composites. This criterion is particularly useful for predicting matrix cracking and fiber breakage, considering the specific fiber orientations and the anisotropic nature of composite materials.

Other criteria specific to composite materials, such as the Hashin and Hoffman criteria, offer unique approaches for modeling failure based on the materials' specific characteristics and loading conditions. These failure models and criteria are integrated into numerical simulations, often via the Finite Element Modeling (FEM), to provide detailed and accurate predictions of the tanks' behavior under various loading and damage conditions.

By combining damage mechanism modeling with appropriate failure criteria, engineers can develop Type IV compressed hydrogen tanks that not only withstand high internal pressures but also maintain their structural integrity over time. This approach ensures the safety, reliability, and longevity of hydrogen storage systems, contributing to the broader adoption of this technology in industrial and transportation applications.

4.10.3.5 Experimental Approaches

One of the most important experimental approaches in the characterization of hydrogen tanks is **pressurization tests** for understanding the mechanical behavior of Type IV compressed hydrogen tanks. These tests involve subjecting the tanks to increasing internal pressures until failure. The primary goal is to measure the maximum strength of the tanks and determine the failure modes, providing crucial data for validating theoretical and numerical models. The tanks are gradually pressurized using appropriate fluids, often water or an inert gas, to minimize explosion risks. This controlled pressurization allows for observing the materials' behavior under increasing stress conditions and identifying potential weak points. By slowly increasing the pressure, it is possible to identify the threshold at which permanent deformations and damage appear, as well as the final rupture point. To monitor the internal conditions of the tank during these tests, pressure and deformation sensors are used. Pressure sensors, placed inside and outside the tank, allow for precise measurement of the internal pressure applied to the tank. These data are crucial for correlating pressure levels to the observed points of deformation and rupture. Simultaneously, strain gauges (extensometers) are installed on the tank's surface to measure the local and global deformations of the composite material. These gauges provide detailed information on the stress and deformation distribution, revealing the areas where the fibers and matrix are most stressed. The data collected from pressure and deformation sensors are then analyzed to understand the failure mechanisms. For instance, deformation variations can indicate the onset of matrix cracking or delamination, while sudden changes in pressure readings may signal imminent rupture. By combining these observations with visual inspections and postmortem characterization techniques such as scanning electron microscopy (SEM) or tomography, researchers can identify specific failure modes, such as fiber breakage, matrix cracking, or delamination. Pressurization tests thus provide essential information for improving the design of Type IV compressed hydrogen tanks. By identifying failure modes and critical pressure levels, engineers can optimize the composite material configuration, improve manufacturing methods, and introduce additional safety measures. These experimental tests, in conjunction with theoretical and numerical models, play a central role in validating tank performance and ensuring their reliability and safety in real-world applications.

In the study of the behavior of Type IV compressed hydrogen tanks, **mechanical testing** on materials is crucial for understanding and quantifying the properties of the composites used. These tests determine parameters such as tensile, compressive, and flexural strength, as well as long-term durability under repeated loads. Tensile, compressive, and flexural tests are conducted on representative samples of the composite materials. Tensile tests measure ultimate strength, elastic modulus, and failure strain, which are essential for understanding the materials' response to common tensile forces in pressurized tanks. Compressive tests provide information on compressive strength and modulus, critical for evaluating stability under compressive loads. Flexural tests determine flexural strength and modulus, important for assessing stiffness and resistance to bending forces.

Additionally, fatigue tests evaluate the materials' durability under repeated load cycles, simulating the filling and emptying cycles of the tanks. These tests determine the fatigue limit, which is the maximum stress level that can be sustained without failure after a specified number of cycles.

The data from tensile, compressive, flexural, and fatigue tests are essential for modeling and designing the tanks. They allow for calibrating theoretical and numerical models, ensuring simulations accurately reflect the materials' real behavior. Moreover, these data help identify weak points and improve the performance of composite materials by optimizing their composition or manufacturing processes.

In the study of the mechanical behavior of Type IV compressed hydrogen tanks, **deformation monitoring and measurement techniques** play an essential role in understanding how materials and structures respond to internal stresses. These techniques allow for real-time monitoring of local and global deformations, as well as the detection of internal defects and structural damage without damaging the tank. The use of strain gauges, digital image correlation (DIC), and fiber optic sensors provides a direct approach for measuring deformations. Strain gauges, also known as extensometers, are devices attached to the tank's surface that measure length changes due to applied stresses. They provide precise data on local deformations at different points on the tank, helping to identify areas subjected to high stress concentrations.

DIC is a nonintrusive optical technique that uses high-resolution cameras to capture a series of images of the tank's surface under different loading conditions. By comparing these images, DIC calculates displacement and deformation fields in great detail. This method is particularly useful for obtaining an overall view of global deformations and detecting areas where damage may develop.

Fiber optic sensors, such as Fiber Bragg Gratings (FBG), are embedded in the composite materials of the tank and measure deformations by altering the properties of light passing through the fiber. These sensors are highly sensitive and can provide continuous, real-time data on internal and external deformations of the tank. Their ability to monitor large areas with high precision makes them a valuable tool for managing the structural integrity of tanks.

In addition to these deformation measurement techniques, nondestructive testing (NDT) methods are used to detect internal defects and structural damage without causing further harm to the tank. Ultrasonics, for example, use sound waves to inspect the interior of materials. Variations in ultrasonic echoes can reveal the presence of cracks, delaminations, or other internal defects. Infrared thermography is another NDT method that measures temperature variations on the tank's surface. Damaged or defective areas often exhibit thermal anomalies due to differences in thermal conductivity. This technique allows for quick detection of potential damage and real-time monitoring of the tank's condition. X-rays, particularly X-ray tomography (CT scan), provide a detailed view of the interior of composite materials, revealing complex internal structures and defects such as porosity, cracks, and foreign inclusions. This method is especially useful for postmortem analyses of failed tanks, enabling a deep understanding of the failure mechanisms. By combining these deformation measurement techniques and NDT methods, researchers and engineers can obtain a comprehensive picture of the mechanical behavior of Type IV compressed hydrogen tanks. This not only improves the design and manufacturing of the tanks but also ensures their safety and reliability throughout their lifecycle.

The experimental approach of **thermal and environmental compatibility testing** is necessary for evaluating the behavior of Type IV compressed hydrogen tanks under various operational conditions. These tanks, often used in variable and sometimes extreme environments, must maintain their integrity and performance despite temperature variations and exposure to corrosive chemical agents.

Thermal testing involves subjecting the tanks to temperature cycles to simulate the conditions they will encounter during use. This includes tests at high and low temperatures, as well as thermal cycles where the tanks are exposed to rapid temperature changes. These tests observe how composite materials react to thermal stresses, which can cause differential expansions and contractions between different material layers. These expansions and contractions can generate internal stresses, potentially leading to matrix cracking, delamination, or other forms of damage.

In addition to thermal cycles, tanks are also exposed to corrosive environments to evaluate their chemical resistance and durability. Composite materials, while generally resistant to corrosion, can be affected by chemical agents present in some industrial or natural environments. Immersion tests in corrosive solutions, such as acids, bases, or salts, simulate these conditions and observe the effects on the materials' mechanical properties. These tests are essential for identifying potential degradations, such as matrix bond weakening or fiber corrosion, that could compromise the tank's integrity.

The study of the effects of temperature variations and chemical agents on the mechanical properties and durability of composite materials also involves mechanical testing after exposure. For example, after subjecting the tanks to thermal cycles or corrosive environments, tensile, compressive, and flexural tests are conducted to measure changes in mechanical properties. Additionally, fatigue tests can be performed to evaluate how these environmental conditions affect the long-term strength of materials under repeated cyclic loads.

These tests provide essential data on dimensional stability, mechanical strength, and the durability of composite materials under realistic conditions. By understanding how materials react to these conditions, engineers can improve tank design to better withstand varied operational environments. This may include selecting resins and fibers more resistant to heat and corrosion, as well as optimizing manufacturing processes to improve cohesion between different material layers. So, thermal and environmental compatibility tests are fundamental for ensuring that Type IV compressed hydrogen tanks maintain their performance and safety under real-world operational conditions. By identifying weak points and improving materials and designs, these tests contribute to the reliability and longevity of the tanks, ensuring safe and efficient use in various industrial and transportation applications.

REFERENCES

1. Mehr, A.S., et al., *Recent challenges and development of technical and technoeconomic aspects for hydrogen storage, insights at different scales; A state of art review.* International Journal of Hydrogen Energy, 2024. **70**: p. 786–815.
2. Azam, M.U., A. Vete, and W. Afzal, *Process simulation and life cycle assessment of waste plastics: a comparison of pyrolysis and hydrocracking.* Molecules, 2022. **27**(22): p. 8084.
3. Palma, V., et al., *A review about the recent advances in selected nonthermal plasma assisted solid–gas phase chemical processes.* Nanomaterials, 2020. **10**(8): p. 1596.
4. Deka, T.J., et al., *Methanol fuel production, utilization, and techno-economy: a review.* Environmental Chemistry Letters, 2022. **20**(6): p. 3525–3554.
5. Brun, K. and T.C. Allison, *Machinery and energy systems for the Hydrogen Economy.* 2022: Elsevier.
6. Olabi, A.G. and E.T. Sayed, *Developments in Hydrogen Fuel Cells.* 2023, MDPI. p. 2431.
7. Greene, J.P., *Automotive plastics and composites: materials and processing.* 2021: William Andrew.
8. Kent, R., *Energy management in plastics processing: strategies, targets, techniques, and tools.* 2018: Elsevier.
9. Shotton-Gale, N., et al., *Clean and environmentally friendly wet-filament winding,* in *Management, recycling and reuse of waste composites.* 2010, Elsevier. p. 331–368.
10. Emonts, C., et al., *Innovation in 3D braiding technology and its applications.* Textiles, 2021. **1**(2): p. 185–205.

5 The Last Word

As we explore the promising future of hydrogen, it is clear that hydrogen holds immense potential in the quest for sustainable and clean energy solutions. This book has meticulously detailed the multifaceted aspects of hydrogen technology, emphasizing its critical role in the global transition toward a greener energy landscape. Hydrogen, often hailed as the fuel of the future, is emerging as a cornerstone of the energy transition. Its versatility as an energy carrier, combined with its potential to produce zero emissions, positions it uniquely to address the pressing challenges of climate change and energy security. Hydrogen's ability to be produced from a variety of resources, including water, natural gas, and biomass, adds to its appeal as a sustainable energy solution.

The environmental benefits of hydrogen are significant. When used in fuel cells, hydrogen combines with oxygen to produce electricity, with water vapor as the only byproduct. This process is not only clean but also highly efficient. Moreover, hydrogen can be stored and transported, making it a flexible energy source that can be used in various applications, from powering vehicles to providing backup power for industries and homes. The potential of hydrogen to significantly reduce GHG emissions makes it a vital component in the fight against climate change. By integrating hydrogen into our energy systems, we can achieve substantial reductions in carbon emissions, contributing to global efforts to limit the rise in average temperatures and mitigate the impacts of climate change.

Throughout the chapters, we have presented the technical, economic, and environmental dimensions of hydrogen energy. Here are some of the critical insights and advancements covered: hydrogen production technologies include electrolysis, SMR, and biological methods. Electrolysis uses electricity to split water into hydrogen and oxygen and is particularly promising when powered by renewable energy sources, offering a pathway to produce green hydrogen. This method is essential for producing hydrogen with minimal environmental impact, especially when coupled with renewable energy sources such as wind and solar power. SMR, currently the most common method for hydrogen production, can be coupled with CCS to reduce its carbon footprint. This integration is crucial for minimizing the environmental impact of hydrogen production from fossil fuels. Emerging techniques using microorganisms to produce hydrogen from organic materials hold potential for sustainable and low-carbon production. These biological methods are innovative and represent the next frontier in hydrogen production, potentially offering more environmentally friendly alternatives.

Hydrogen storage solutions are diverse and include compressed hydrogen, liquid hydrogen, and solid-state storage. Storing hydrogen in high-pressure tanks is a mature technology suitable for various applications, including fuel cell vehicles. This method ensures that hydrogen can be efficiently stored and utilized when needed.

DOI: 10.1201/9781032718453-5

"""

Cryogenic storage of hydrogen in liquid form is essential for certain industrial applications and long-term storage. This method is vital for applications requiring large volumes of hydrogen, as it significantly reduces the space needed for storage. Advanced materials like metal hydrides and carbon nanotubes offer innovative ways to store hydrogen efficiently and safely. These materials enable the storage of hydrogen at lower pressures and temperatures, enhancing safety and efficiency.

Hydrogen applications and integration span transportation, industry, and energy storage. Hydrogen fuel cell vehicles are gaining traction as a clean alternative to ICEs, with several automakers investing heavily in this technology. These vehicles produce zero emissions at the tailpipe, making them an attractive option for reducing air pollution and GHG emissions. Hydrogen is used in various industrial processes, including refining, ammonia production, and metal processing, offering a pathway to decarbonize these sectors. The integration of hydrogen into these processes is crucial for achieving significant reductions in industrial carbon emissions. Hydrogen can store excess renewable energy, providing a buffer against intermittency and enhancing grid stability. This capability is vital for balancing supply and demand in energy systems increasingly reliant on variable renewable energy sources.

The transition to a hydrogen economy is not without challenges. Significant investments are needed in infrastructure, such as hydrogen refueling stations, pipelines, and storage facilities. Additionally, regulatory frameworks must be developed to ensure the safe and efficient deployment of hydrogen technologies. These investments and regulations are critical for building a robust hydrogen infrastructure that supports widespread adoption. Research and development are crucial to overcoming the technical and economic barriers associated with hydrogen production, storage, and utilization. Innovations in electrolyzer efficiency, hydrogen storage materials, and fuel cell technology are essential to make hydrogen a competitive energy carrier. These advancements will help reduce costs and improve the performance of hydrogen technologies, making them more attractive to consumers and businesses. International collaboration will play a vital role in this transition. The establishment of global partnerships and networks can facilitate the sharing of knowledge, technology, and best practices, accelerating the adoption of hydrogen solutions worldwide. Collaborative efforts will help overcome common challenges and promote the development of a global hydrogen economy.

Looking ahead, the path to a hydrogen-powered world is filled with opportunities for innovation and growth. Hydrogen stands at the forefront of the energy transition, offering a versatile and sustainable solution to some of the most pressing energy challenges of our time. This book has endeavored to provide a comprehensive overview of the current state of hydrogen technology and its potential to transform the energy landscape. As we move forward, it is imperative to continue investing in and supporting the development of hydrogen technologies. Governments, industries, and research institutions must work together to create an enabling environment that fosters innovation, reduces costs, and promotes the widespread adoption of hydrogen.

As we look to the future, the continued development and deployment of hydrogen technologies will be instrumental in achieving a low-carbon economy. The journey toward a hydrogen-powered world is just beginning, and with concerted effort and innovation, hydrogen can indeed become a pivotal element in the global energy mix.

The insights and advancements presented in this book serve as a testament to the incredible potential of hydrogen. By harnessing this potential, we can pave the way for a cleaner, more sustainable future for generations to come. Let us embrace the challenge and opportunity that hydrogen presents, and work together to build a brighter, greener future for all.

Comprehensive Glossary of Hydrogen Technology Terms and Definitions

Absorption: The process by which a substance is taken up by another substance, such as hydrogen being absorbed by metal hydrides for storage.

Absorption Chillers: Cooling systems that use heat (often from waste energy or solar power) to drive a refrigeration cycle, which can be paired with hydrogen systems for efficiency.

Adsorption: The process in which atoms, ions, or molecules from a gas, liquid, or dissolved solid adhere to a surface of a solid or liquid.

Advanced Composite Materials: High-strength, lightweight materials are used in the construction of hydrogen storage tanks and other components.

Advanced Electrolyzers: New generations of electrolyzers that offer higher efficiency and lower costs for hydrogen production from water.

Ammonia (NH_3): A compound of nitrogen and hydrogen used as a hydrogen carrier due to its high hydrogen content and potential for carbon-free production when using renewable energy.

Ammonia Cracking: A process to decompose ammonia into hydrogen and nitrogen, used for hydrogen storage and transport.

Artificial Intelligence (AI): The simulation of human intelligence processes by machines, which can enhance the safety and efficiency of hydrogen technologies.

Biological Hydrogen Production (Biohydrogen): The generation of hydrogen by microorganisms such as bacteria and algae through biological processes like fermentation.

Biohydrogen Production: Generating hydrogen from biological sources and processes, such as algae or fermentation of organic materials.

Biomass Gasification: A process that converts organic materials into hydrogen, carbon monoxide, and carbon dioxide by reacting the materials at high temperatures without combustion.

Blue Hydrogen: Hydrogen produced from natural gas with carbon capture and storage (CCS) to reduce carbon emissions.

Borohydrides: Chemical compounds consisting of boron and hydrogen are used in hydrogen storage due to their high hydrogen content.

Carbon Capture and Storage (CCS): A technology to capture carbon dioxide emissions from sources like power plants and store it underground to prevent it from entering the atmosphere.

Carbon Dioxide Hydrogenation: A chemical reaction that converts carbon dioxide and hydrogen into hydrocarbons, often used in synthetic fuel production.

Carbon Nanotubes: Cylindrical molecules with novel properties, used in hydrogen storage due to their high surface area and adsorption capacity.

Catalyst Poisoning: The deactivation of a catalyst due to the presence of impurities, a concern in hydrogen production and fuel cell technologies.

Catalysis: The acceleration of a chemical reaction by a catalyst, essential in hydrogen production processes such as steam methane reforming.

Circular Economy: An economic system aimed at eliminating waste and the continual use of resources, with hydrogen playing a role in reducing emissions and recycling materials.

Compressed Gas Storage: A method of storing hydrogen at high pressures in specially designed tanks.

Compressed Hydrogen Storage: Storing hydrogen in a compressed gaseous form, often at pressures up to 700 bar for fuel cell vehicles.

Cryo-Compressed Hydrogen Storage: A combination of cryogenic and compressed hydrogen storage methods, achieving higher density and efficiency.

Cryogenic Storage: The storage of hydrogen in liquid form at extremely low temperatures, significantly reducing its volume for storage and transportation.

Desorption: The process by which a substance is released from or through a surface, opposite of adsorption.

Diffusion Phenomena: The movement of hydrogen molecules through materials, which is a key consideration in storage and transport.

Direct Methane to Hydrogen (DMH): A process to produce hydrogen directly from methane without intermediate steps, aiming for higher efficiency and lower emissions.

Distributed Hydrogen Production: Producing hydrogen close to where it will be used, reducing transportation and infrastructure costs.

Electrochemical Compression: Using electrochemical cells to compress hydrogen to high pressures, enhancing storage efficiency.

Electrochemical Hydrogen Compression: Using electrochemical cells to compress hydrogen gas to high pressures, improving storage efficiency.

Electrochemical Methanol Decomposition: A method of producing hydrogen by decomposing methanol through an electrochemical process.

Electrolysis: A process that uses electricity to split water into hydrogen and oxygen, especially effective when powered by renewable energy sources to produce green hydrogen.

Environmental Impacts: The effects hydrogen production, storage, and utilization have on the environment, including emissions and resource use.

Fuel Cell: A device that converts the chemical energy of hydrogen into electricity, with water vapor as the only byproduct.

Fuel Cell Efficiency: The ratio of the electrical energy output to the chemical energy input in a fuel cell, an important performance metric.

Fuel Cell Vehicles (FCVs): Vehicles that use fuel cells to convert hydrogen into electricity to power electric motors, offering zero emissions.

Gas Diffusion Layer (GDL): A component of fuel cells that facilitates the transport of reactants to the catalyst layers and the removal of water and heat.

Graphene: A form of carbon consisting of a single layer of atoms, with high potential for hydrogen storage due to its large surface area.

Green Hydrogen: Hydrogen produced using renewable energy sources, resulting in minimal environmental impact.

High-Pressure Electrolysis: A method of electrolysis conducted at high pressures to increase the efficiency of hydrogen production.

Hydride Formation Reactions: Chemical reactions where hydrogen is absorbed into a material to form a hydride, used in hydrogen storage.

Hydrogen Blending: The process of mixing hydrogen with natural gas for use in existing natural gas pipelines and infrastructure to reduce carbon emissions.

Hydrogen Carriers: Substances or materials used to transport hydrogen, including liquid organic hydrogen carriers and metal hydrides.

Hydrogen Economy: An economy based on hydrogen as a major energy carrier, envisioned to reduce carbon emissions and reliance on fossil fuels.

Hydrogen Embrittlement: The process by which metals become brittle and fracture due to the absorption of hydrogen, a concern in pipeline transport.

Hydrogen Fueling Infrastructure: The network of production, storage, and dispensing facilities needed to support hydrogen fuel cell vehicles.

Hydrogen Fueling Stations: Infrastructure for refueling hydrogen-powered vehicles, including storage, compression, and dispensing systems.

Hydrogen Leak Detection: Technologies and methods used to detect hydrogen leaks, essential for safety in hydrogen systems.

Hydrogen Life Cycle: The entire process of hydrogen production, storage, distribution, and utilization, including its environmental impacts.

Hydrogen Mobility: The use of hydrogen as a fuel for transportation, including fuel cell vehicles and infrastructure like refueling stations.

Hydrogen Peroxide Decomposition: A method of producing hydrogen by decomposing hydrogen peroxide.

Hydrogen Production: Various methods of producing hydrogen, including electrolysis, steam methane reforming, and biological processes.

Hydrogen Purification: The process of removing impurities from hydrogen to meet quality standards for various applications, including fuel cells.

Hydrogen Sensors: Devices used to detect the presence of hydrogen gas to ensure safety in hydrogen production, storage, and utilization.

Hydrogen Storage: Various methods of storing hydrogen, including compressed gas, liquid hydrogen, and solid-state storage.

Hydrogen Technology: Technologies involved in the production, storage, and utilization of hydrogen as an energy carrier.

Hydrogen Valleys: Regions or areas where hydrogen production, storage, and utilization are integrated into the local economy and infrastructure.

Hydrogenation: A chemical reaction between hydrogen and another compound, often used in industrial processes.

Hydrothermal Gasification: A process that converts organic material into hydrogen-rich gas using high temperature and pressure water, without the need for drying the feedstock.

Industrial Hydrogen Applications: The use of hydrogen in various industrial processes such as refining, ammonia production, and steelmaking.

ISO 16111: An international standard for the safety of hydrogen storage systems using metal hydrides.

ISO 19880-1: An international standard for the design, construction, and operation of hydrogen fueling stations.

Liquid Organic Hydrogen Carriers (LOHCs): Organic compounds that can absorb and release hydrogen through chemical reactions, used in hydrogen storage and transport.

Membrane Electrode Assembly (MEA): A key component of fuel cells, consisting of a membrane and electrodes where the electrochemical reactions occur.

Metal Hydrides: Compounds formed by hydrogen and metals, used for hydrogen storage due to their ability to absorb and release hydrogen.

Methanolysis: A chemical process that produces hydrogen from methanol.

Microgrids: Small-scale power grids that can operate independently or in conjunction with the main electrical grid, often using hydrogen for energy storage and management.

Nanomaterials for Hydrogen Storage: Advanced materials with nanoscale structures that enhance hydrogen storage capacity and kinetics.

Natural Hydrogen Production: Hydrogen that occurs naturally in geological formations, which can be harnessed for energy use.

NFPA 2: A code from the National Fire Protection Association that provides safety requirements for hydrogen technologies.

NFPA 55: A code from the National Fire Protection Association that addresses the storage, use, and handling of compressed gases, including hydrogen.

Offshore Hydrogen Production: Generating hydrogen from offshore wind or solar energy sources, often using electrolysis platforms located at sea.

Onboard Hydrogen Storage: Storage systems integrated into vehicles to hold hydrogen fuel for fuel cell or internal combustion engine use.

Phase Changes and Structural Transitions: Changes in the physical state of hydrogen storage materials that affect their performance and storage capacity.

Photoelectrolysis: A process that uses light energy to split water into hydrogen and oxygen, a method for producing green hydrogen.

Plasma Reforming: A hydrogen production method using plasma technology to convert hydrocarbons into hydrogen and carbon monoxide.

Polymer Electrolyte Membrane (PEM): A type of membrane used in fuel cells and electrolyzers that conducts protons while being impermeable to gases.

Power-to-Gas (P2G): A technology that converts electrical energy into hydrogen or synthetic natural gas for storage and use in the gas grid.

Proton Exchange Membrane Fuel Cells (PEMFC): Fuel cells that use a polymer electrolyte membrane to conduct protons and generate electricity from hydrogen.

Pyrolysis: A thermochemical decomposition process of organic material at elevated temperatures in the absence of oxygen, used in hydrogen production.

Renewable Energy Sources: Energy sources that are replenished naturally, such as solar, wind, and hydroelectric power, which can be used to produce green hydrogen.

Reversible Fuel Cells: Fuel cells that can operate in both directions, generating electricity from hydrogen and producing hydrogen from electricity.

Safety Standards: Regulations and guidelines to ensure the safe production, storage, and utilization of hydrogen.

Solid Oxide Electrolysis Cells (SOEC): Electrolyzers that use a solid oxide or ceramic electrolyte to split water into hydrogen and oxygen at high temperatures.

Stationary Hydrogen Applications: The use of hydrogen in nonmobile applications such as backup power systems, residential energy storage, and industrial processes.

Steam Methane Reforming (SMR): A method of producing hydrogen from natural gas by reacting it with steam, currently the most common hydrogen production method.

Storage Tanks: Containers used to store hydrogen, either in compressed gas, liquid, or solid form.

Synthetic Natural Gas (SNG): A fuel produced by methanation of hydrogen and carbon dioxide, used as a substitute for natural gas.

Thermochemical Decomposition of Water: A high-temperature process to produce hydrogen by breaking down water molecules.

Thermochemical Water Splitting: A process that uses high temperatures to split water into hydrogen and oxygen, often driven by solar or nuclear heat sources.

Transport Hydrogen Applications: The use of hydrogen as a fuel for various modes of transportation, including cars, buses, trucks, trains, ships, and aircraft.

Type IV Compressed Hydrogen Tanks: Advanced storage tanks made of composite materials designed to store hydrogen at high pressures safely.

Underground Hydrogen Pipelines: Pipelines used for transporting hydrogen over long distances, often buried underground for safety and security.

Underground Hydrogen Storage: Storing hydrogen in underground caverns or depleted oil and gas fields for large-scale and long-term storage.

Vapor Pressure: The pressure exerted by a vapor in thermodynamic equilibrium with its condensed phases at a given temperature, relevant in hydrogen storage and transport.

Water Electrolysis: The process of using electricity to split water into hydrogen and oxygen, a primary method for producing green hydrogen.

Water–Gas Shift Reaction: A reaction in which carbon monoxide and water vapor are converted into carbon dioxide and hydrogen, used in hydrogen production.

Wind-to-Hydrogen: Generating hydrogen from wind power using electrolysis, a renewable and sustainable method for hydrogen production.

Zero-Carbon Hydrogen: Hydrogen produced without any associated carbon emissions, typically using renewable energy sources.

Zero-Emission: A process or product that does not release any pollutants or greenhouse gases into the atmosphere, as aimed for with hydrogen technologies.

R

raw materials, 66
reducing risks, 31
regulatory pathway to China's hydrogen economy, 35
renewable energy, 12
rotational molding, 198

S

safety, 25
safety standards, 26
security and logistics, 80
single-walled carbon nanotubes (SWCNTs), 145
SMR, 12, 19, 47, 48, 63, 67, 78
SOECs, 82
solid material adsorption tanks, 156
solid oxide electrolysis cell (SOEC), 72, 82
solubility, 215
sorption models, 127
source of electricity, 15
stainless steel tanks, 152
steam methane reforming (SMR), 6
steam reforming, 63
STEM education, 32

T

technical and economic challenges, 56
thermal phenomena, 117
thermomechanical decomposition of ammonia, 101
thermomechanical decomposition of water, 102
thermomechanical water splitting, 81
transport, 15

U

US hydrogen policy, 34

V

vacuum insulation tanks (VLIs), 154

W

water electrolysis, 71, 74
water photoelectrolysis, 104
water-gas shift (WGS), 64
WGS, 65, 67
World Bank Group, 10